A TIMETABLE
OF INVENTIONS
AND DISCOVERIES

By the same author

The world water speed record
with Leo Villa 1976

The Guinness guide to water skiing
with David Nations 1977

The Guinness book of motorboating
facts and feats *1979*

Life with the speed king *1979*

En l'an 200 *with Alex Albani 1981*

The man with two shadows *1981*

Granada Guide Series –
Racing Cars *1981*

Richard Shuttleworth –
an illustrated biography *1982*

A TIMETABLE
OF INVENTIONS
AND DISCOVERIES

from pre-history to the present day

Kevin Desmond

M. Evans & Company, Inc.   New York

Originally published in Great Britain in 1986 by Constable and
Company Limited

M. Evans and Company, Inc.
216 East 49 Street
New York, New York 10017

ISBN 0-87131-520-3

Manufactured in the United States of America

9 8 7 6 5 4 3 2 1

This chronology is dedicated to GOD Creator of all things

God hath made man upright; but they have sought out many inventions.

Eccleslastes vii 9, c. 250 BC

Countless ages will beget many new inventions, but my own is mine.

Pedanius Dioscorides, *fl.* first Century AD

Tell me when anything was ever made.

Leonardo da Vinci, *c.* 1515

'Tis frivolous to fix pedantically the date of a particular invention. They have all been invented over and over fifty times. Man is the arch machine of which all these shifts drawn from himself are toy models.

Ralph Waldo Emerson, *Conduct of Life: Fate,* 1860

General thanks are due to the public relations departments of the many companies whose innovations appear in this book – too numerous to single out; thanks also to those editors who published my 'appeals' in their magazines – and to the choice few who responded to them. I would also have been lost but for the patient help of a whole chain of librarians – particularly those in my own locality of North-West London.

Last but not least, special gratitude must go to Andrew Macdonald who has done such fine additional illustrations; to Keith Macqueen for a mammoth indexing task; to Imogen Olsen; to my editors Miles Huddleston and Prudence Fay; to Ivor Kamlish for the typography and design; but above all to my long-suffering wife, Alex.

K.D. 1986

Acknowledgements

The scientific and industrial progress recorded in this book has been achieved in an extraordinarily short period in history. Scientists tell us that the earth is about 4½ billion years old. Imagine this period represented by a 24-hour earth clock. The first faint traces of life appeared at about 2 p.m. (1400 hours). The dinosaur showed up at about 11 p.m. (2300 hours). And the human species? Man finally made the scene at two seconds before midnight. The entire last 6000 years of human endeavour have occurred in the final one-tenth of a second.

Within that tenth of a second, it took man 1750 years from the year 1 AD to double his technological knowledge. A mere 150 years later, by the year 1900, he had doubled his knowledge again. The next doubling took only 50 years (1900–50); the next, ten years (1950–60); and man's knowledge has quadrupled during each of the last two decades.

Another way of estimating the same progress is to state that there have been somewhere between 40,000 and 60,000 generations of *Homo sapiens* on this planet. In Great Britain, where the Industrial Revolution was born, only seven generations have been involved with that revolution – just a blip in history.

If most inventors' lives were to be averaged out, one would find that they spent 23 out of their 70 full years fast asleep, that they spent six years eating and two years keeping themselves clean. Given that no fewer than twenty of the years remaining to them are merely formative – devoted to the learning of established science – this leaves them just twenty years of their entire life to accomplish the practical work of inventing or discovering something totally new. No wonder the era of the individual inventor is no longer with us and has been replaced by ever-growing Research and Development armies of university-linked, multi-national corporations.

One should not leave out the fact that *only half* of the world's total accumulated population throughout history was given the opportunity to invent or discover. The women who feature in this chronology may be counted on the fingers of both hands. Only now perhaps is a 6000-year-old prejudice being overturned: that woman's only role has been to conceive and bring up the latest generation to both wield and suffer from the newest weaponry which man has been able to devise.

A major part of our recent and astonishing development of technological knowledge has been due to the use of an essentially simple tool – the computer. It has been projected that, by the end of this millennium, artificially intelligent computers will have outpaced human intelligence. Yet can this be possible? The human brain can store about 2½ million times more information than today's most advanced computer. The human brain can hold some 10 million million 'bits' of information, and all tucked away in a skull that measures about one-twentieth of a cubic foot. The human brain weighs about 3 lb. A modern mini-computer – a 1000 million bit machine – takes up about 40 cubic feet and holds about 20 lb of memory units. To attempt to equal – let alone surpass – *cerebrum humanum* would probably require the long-term pooling of all the available computer scientists, psychologists, linguists and brain surgeons in the USSR, USA, Europe and Japan.

Writing in severely impoverished conditions in 1919, William Friese-Greene, English inventor of the cinema some 40 years earlier, recalled how '. . . the expense and enthusiasm of the Inception and for proving the possibilities of Cinematography, ended my business career. One man cannot compete against the prejudice and jealousy of an Invention which is not believed at the time. So the Invention goes through the stages: 1st: Rubbish, impossible! 2nd: Impossible! 3rd: Possible and probable. 4th: Yes, possible, but is there any money in it? 5th: Yes there is money in it (that is the stage of worry). 6th: Plenty of money in it. 7th: Oh I have heard of this Invention long ago! So the ebb and flow of Invention, like the ebb and flow of Reason, will always flow from generation to generation . . .'

Almost endorsing Friese-Greene's analysis, there is now such a library of science-fiction books and films that whenever an item moves out of the realm of fiction and into that of fact – in other words, is *really* invented or discovered – people are so well prepared for it that the reality no longer awes or impresses them. They too have heard of it long ago.

Although three separate topics appear in the title of this book, there are no clear-cut divisions between them. The process is too fluid for that. An 'invention' often appears as a more noticeable step forward than an 'innovation', the latter being defined as a new assembly of existing inventions. 'Discoveries' are usually made either by the invention of new devices or by an innovative assembly of existing equipment with which to observe, measure and demonstrate. Once made, such 'discoveries' in their turn may be utilised as just one part in the innovative assembly that makes up a distinguishable 'invention'.

Books about the history of invention usually choose to list them in alphabetical order or grouped by subjects, appending a chronology of no more than a couple of dozen pages at the back. They also usually give much space to illustrating and explaining how the inventions worked and work. This in turn, for reasons of space, obliges them to limit the nature of the inventions listed to practical day-to-day objects.

In this book I have taken a different approach. As almost 4000 entries have been compiled, it has not been possible to carry out conclusive verifications on the originality or 'first-ness' of each and every entry, but the new chronological approach has enabled me to include a wider range of topics – both the purely scientific and the more whimsical and forgotten. This book therefore does not seek to rival the more conventional approach, but rather to act as a companion to it.

It has been estimated that there is now an accumulation of 25 million patent documents worldwide, including microfiche and ultrafiche systems. Well over $4\frac{1}{2}$ million of these are official US patents, with a total of over 2 million for both the UK and France, not to mention totals of 500,000 to over a million for other industrial nations. Even as you read them, such statistics are already out of date.

Currently anything from 600,000 to three-quarters of a million *new* patent specifications worldwide are accepted each year. In recent years, from 28,000 to 38,000 patents have been granted annually by the UK Patents Office alone – which means that over 500 new patents are laid out for public inspection at Southampton Buildings, London, each week.

As was seen at the start of this Introduction, the rate at which man is inventing and discovering has speeded up tremendously in the past few decades. (Future books of this kind will show whether this frenetic trend has been maintained.) The entries which follow, therefore, can only represent a minute and symbolic fraction of man's astounding technological ingenuity.

Despite that, listing inventions and discoveries in chronological order allows the reader to chart the pattern of technological progress in a slightly different manner: to see, for instance, how certain basic inventions were the seed for a vast plantation of subsequent applications. Witness as such the toothed wheel, the steam engine, steel, the internal-combustion engine, plastic, the thermionic valve, the transistor and the micro-processor.

A year-by-year account will also enable the family historian, for example, to see exactly what were the innovations that would have excited each generation of his ancestors, to find out what was being discovered in the year when he was born, and finally, using the list of inventions still to be realised, to glimpse what may become the latest reality for his children, and their children.

So what direction will man's ingenuity take from now on?

In the course of the past forty years, half a person's lifetime, we have, through invention and innovation, built up an arsenal of weaponry capable of destroying the earth several times over. We have also so 'organised' our trade and industrial environment that the vulnerable wildlife kingdom, which has taken millions of years to evolve, could be virtually wiped out by the end of the century.

And yet in that same short period, we have come up with both cures and preventions for many illnesses hitherto considered fatal; we can replace many parts of the human anatomy that are damaged or in danger of breakdown; we can in a few minutes perform mathematical computations which would have once taken teams of workers years to complete; we have at last begun to quantify the real links between the innermost secrets of the elusive atom and the outmost mysteries of the universe; finally, we have so improved our telecommunication systems that the world, and even the ionosphere above it, acts as a 'global village', able to show immediate international concern for any disaster which may occur in it.

In the 1890s, Charles Duell, the then US Commissioner of Patents and Trademarks, advocated the abolition of his office on the grounds that everything worthwhile had already been invented. He encountered polite derision.

It would be about as impossible to use a laser beam to change the flight-path of a pre-historic flint spearhead without shattering it, as to change either the pace or the direction of invention without killing it dead.

At the last count, it may only be our deep instinct for survival coupled with that remarkable ingenuity which is evidenced in the following pages, that can give us cause to look bravely ahead towards an even more dazzling, but perhaps less destructive, scientific and technological future.

Kevin Desmond, London
7 April 1985

15,000 million years ago
Formation of the universe ('Big
Bang').

4600 million years ago
Formation of our solar system.

Pre-dates (estimated)

3,000,000 years ago
Flint-knapping.

1,800,000 years ago
Stone tool kits (deliberate
fracture)
Koobi Fora, Rift Valley East Africa.

1,500,000 years ago
Homo erectus.

250,000 years ago
Homo sapiens.

230,000 years ago
Hand-axes, cleavers and flake
tools (stone)
*'Acheulean' industry from the
Kapthurin beds near Lake Baringo,
Kenya, Africa.*

107,000 years ago
Engraving (on bone)
Peche de l'Aze, Dordogne, France.

45,000 years ago
Endscrapers, burins, backed
knives (stone and bone)
*Central and Eastern Europe. Used to
scrape hides clean of hair and fat.*

27,000 years ago
Engraving (on stone)
*La Ferrassie, Périgord, France. Large
blocks of stone engraved with animal
figures and symbols.*

27,000 years ago
Sun-dried clay artefacts.

27,000 years ago
Paintings of animal figures on
rock slabs
*at the Apollo II shelter, Namibia,
Africa.*

15,000
Cave painting
Lascaux, France. Depicting reindeer and bulls.

13,000
Cave mural
Altamira, Spain. Depicting bison herds.

12,000
Harpoons, spearthrowers and fishing hooks
Man-made fire.

11,000
Domesticated animal husbandry
Thailand, South-east Asia.
Fired clay objects (ceramic) *animal and human figurines.*
Stone lamps (animal-fat fuel).

10,000
Mortar and pestle and digging stick.
The zero (mathematics)
Hindu priests.

The Aurignacian 'Laussel Venus'

22,000
Sculpture (in stone)
the Aurignacian 'Laussel Venus'.

17,000
Sewing needles (in bone)
Pleistocene peoples of Europe and North Asia.

9–8000
Domestication of crops (wheat and barley)
The Near East.
Flint arrowheads and bows.

8350–7350
Walled town
Jericho, ancient city of Israel.

8050
Domesticated goat
Asiab, Iran.

8000
Quern (grain-crusher)
Ships (reed-boats and shaped dug-outs).

7700
Domesticated dog
North Yorkshire, England.

7200
Domesticated sheep
Thessaly, Greece.

7000
Coiled pottery
Kurdistan, Iran.
Domesticated peas and beans
Domesticated pigs and cattle
Thessaly, Greece.

6500
Potter's wheel
Asia Minor.

6000
Beer (from malted grain)
Mesopotamia, South-west Asia.
Rice cultivation
Thailand.
Rotary quern
Mesopotamia.
Villages (4–5 hectares)
Mesopotamia. With irrigation and religious temples.

5000
Drilling holes by tying rods together with thongs
Flaked-stone hoe blade (agricultural)
Hassana, Mesopotamia.
Pottery repair technique
Nineveh, Northern Mesopotamia.
Woven cloth

4500
Copper-working

earliest smelting sites, Rudna Glava, Yugoslavia.
Stone door sockets (prototype hinges)
Hassana, Mesopotamia.

4000
Bevel-rimmed open bowl (mass-produced)
Mesopotamia.
Copper axes
Plough (copper parts)
Reed ropes

3875
Harp (musical instrument)
attributed to Jubal, *son of Lamech and Adah (see Genesis) – 'father of all such as handle the harp and organ'.*

3800
Maps (clay tablets)
Sumer.

3761
Jewish Year 1.

3500
Cylinder seal
Mesopotamia.
Gold (mined)
Mesopotamia.
Pictographic statement of accounts
at the temple of Uruk, a city-state of Sumer.
Rotary grain grinder
Mesopotamia.
Wheeled carts
Sumer and Syria. Found in tombs at Kish and Ur with copper-nail rim studs.

3200
Ship (rectangular sail)
depicted on pottery vessel from the Egyptian Naqada II period.

3150
Irrigation
River Nile, Upper Egypt. Evidenced

on a ceremonial macehead from the
reign of 'King Scorpion'.

3100
Irrigation (large-scale)
'King Menes' in the Nile Valley,
Egypt.
Pictographic inscription
the slate Palette of Narmer, King of
Upper and Lower Egypt.

3000
Boat (timber-built)
River Nile, Egypt, using ash, meru or
'wood from the Lebanon'.
Warp-weighted looms.

2950
Unwritten papyrus
found in a tomb at Saqqara, Egypt,
from the reign of King Den.

2950–2750
Dam (masonry walls)
Garawi Valley, Egypt.

2800
Bronze working
South-west Asia.

2761
Jewish Year 1000.

2650
Stepped pyramid
Imhotep, Egyptian architect and
vizier for King Djoser at Saqqara.

2640
Silk production
Si-Ling Chi, legitimate wife of
Prince Hoang-ti of China. For her
achievement, Si-Ling Chi was deified.

2600
Monumental hieroglyphic texts
Egyptian tombs.
**Scribe's outfit (palette, bag for
powdered pigments and reed-
holder)**
Egypt.

Ink (ordinary)
Egypt and China. Lampblack ground
with a solution of glue or gums.

2590
Embalmed person
Egyptian Queen Hetepheres, mother
of Cheops. Only her internal organs
have survived.

2575
Smooth-sided pyramid
King Sneferu at North Dahshur,
Egypt.

2568 (?)
Ark (animal rescue ship)
Noah, as directed by God. Constructed
of gopher wood and used during the
Great Flood. The Ark was later
'beached' on the side of a mountain
called Ararat in East Turkey.

2500
Bitumen as glue
Mesopotamia.
Cotton (spun and woven)
India.
Gaming boards
Ur, city-state of Sumer.
**Plough (ox-pulled and man-
guided)**
Egypt.
Stone-working drill
Egypt.

From the Tomb of Rekhmire at Thebes

2450
**Senet (board game similar to
draughts)**
Egypt.

Sistrum

2400
Engraved seals
India.
History in hieroglyphs
the 'Palermo Stone'.
Mummy of 'Waty'
*Found in the tomb of Nefer, Royal
Court singer at Saqqara, Egypt.*
Sistrum (musical rattle)
Egypt.
Written papyrus
*the Abusir Papyrus, from the reign of
the Fifth Dynastic King Djedkare
Isesi.*

2380
'Wisdom Text'
*the instructions and precepts of
Ptahhotep, vizier of King Djedkare
Isesi of Egypt.*

2350
**Lightweight, horse-drawn
chariot (spoked wheels)**
the Hurrians of Mesopotamia.

2300
Artificial stone (glazed)
Telltaya, Mesopotamia.

2200
Rough engraving of city layout
Babylon.

2100
Ur-Nammu Ziggurat (building)
Mesopotamia.
Workable road network
Ur-Nammu of Mesopotamia.

2000
**Chariot wheels (copper-tyred as
opposed to studded)**
Susa, capital of Elam.
Snow skis
*Northern Norway. Depicted in a rock
carving known as 'the Rødøy Man'.*

1900–1800
**Gynaecological/Veterinarian
papyrus**
the Illahun Papyrus.

1800
Bath tub
Mari, Babylonia.
**Geometrical problem texts (on
clay)**
*Tell Harmal (Baghdad). Includes a
demonstration of triangular laws.*

1761
Jewish Year 2000.

1700
**'Hounds and Jackals'
(sophisticated board game)**
Egypt.
The Phaistos Disk
Crete.
Royal bath tub
*found in the Queen's bathroom at the
palace of King Minos in Knossos,
Crete.*

1650–1600
Mathematical papyrus

the 'Rhind Papyrus', a copy of a much older document.

1600
Alphabet (Sinaitic)
Canaanite scholars.

1500
Contraceptive
description of vaginal plug of lint, ground acacia branches and honey in the Egyptian Ebers Papyrus.

Egyptian war chariot

Glass manufacture
Mesopotamia.
Glazing of pottery
Nuzi, Mesopotamia.
Horse bit (bronze with cheek pieces)
Egypt.
Iron smelting
the Hittites of Anatolia.

Bronze sword hilt

Phoenician alphabet
(30-letter system).
Plank-built boats
North Ferriby, beside the River Humber, Yorkshire, England.
Ritual astromical stone clock
Stonehenge, Salisbury Plain, England.

1470
Polychrome glass vessels
during the reign of King Tuthmosis III of Egypt.

1417–1379
The Colossae of Memnon
King Amenhotep III of Egypt. Following damage by an earthquake, whenever the sun heated the air inside the statue, it expanded and rushed through the cracks to make a 'whistling' sound.

1400
Accurate plan of a city
Nippur, Mesopotamia.
Horse-drawn chariots (paired-draught animals)
both Shang Dynasty in China and under Thotmes III and IV in Egypt.
Shaduf (well-sweep with counterpoise)
Egypt

1320
Papyrus map
sketch of an Egyptian gold mine.

1300
Complete alphabet
Ugarit, Syria. 32 letters.
Mobile prefabricated temple
Bezaleel ben Uri (Tribe of Judah), together with Aholiab ben Ahisamach (Tribe of Dan) for Moses. This included the tabernacle, ark, mercy seat, candlestick, priests' garments and all else as specified in Exodus.
Musical notation (on clay tablets)
Ugarit, Syria.

1270
Encyclopaedia
Abulfaraj (Bar-Hebraeus), *Syrian historian. As master of Syriac, Arabic and Greek, and equally learned in philosophy, theology and medicine, Abulfaraj was admirably suited for the task.*
Sonic destructor
Joshua, *son of Nun, during his conquest of Canaan. The blowing of trumpets, shaped like ram's horns, with mass shouting is reported to have destroyed the walls of Jericho.*

1250
The Wooden Horse of Troy (disguised troop carrier)
mythical artifice by which the Greeks are said to have got possession of Troy, with assistance from the captive, Sinon.

1200
Horse-bit (jointed mouthpiece)
Egypt.

1150
Chess (board game)
popularly attributed to Palamedes, 'the handy or contriving one', a mythical figure, son of Nauplius, and also erroneously credited with the invention of lighthouses, measures, scales, the discus, dice, the alphabet and the art of regulating sentinels.

1100
Bell (tintinabulum)
Babylonian Palace of Nimrod.
Dictionary
Pa-out-She, *Chinese scholar. 40,000 characters, mostly hieroglyphic or zodiacal.*
Phoenician alphabet (22-letter system)
Byblos and Ras Shamra, the Lebanon.

1000
Protogeometric style of pottery
Eastern Mediterranean.

Wooden toy tiger (movable jaw)
Thebes, Upper Egypt.

950
Solomon's Temple
Hiram of Tyre, *Phoenician architect.*

Assyrian army buoyancy aids

880
Inflatable swimming aid
King Assur-nasir-apli II of Assyria.
To enable his troops to cross a river.

870
Canal-building
King Assur-nasir-apli II of Assyria.
From off the Upper Zab river to water the orchards and botanic gardens of his 'creation' at Kalhu.

865
Siege engines
Assyrian generals.

850
Battering rams
Assyrian generals.
Biremes (galleys with two banks of oars)
Assyrian boatbuilders.
Damp-proofing (bitumen process)
Assyrian architects.

Stirrup (horses)
Assyrian blacksmiths.

775
Fully phonetic alphabetic script
Greece. Written from left to right.

770
Gold coins (uninscribed)
Chou Dynasty of China.

761
Jewish Year 3000.

750
Triremes (galleys with three banks of oars)
Greek boatbuilders.

730
The Sundial of Ahaz
Ahaz, son of Jotham and eleventh King of Judah.
Written poetry
'The Iliad' by Homer *and 'Works and Days' by* Hesiod, *both Greek poets.*

Greek warrior's helmet, 750 BC

720
Aramaic/Assyrian alphabet.

700
Aqueduct
King Sennacherib of Assyria, *during the rebuilding of Nineveh. Its length was 30 miles.*
Demotic script
Egypt.

673
Lyre (seven-stringed)
Terpander, *possible founder of the first Greek school of music at Sparta.*

650
Glass-making texts (on clay tablets)
Nineveh, Assyria.

640
Roof tiles
on the Temple of Hera, Olympia, Greece.

605
Circus Maximus
Lucius Tarquinius Priscus, *King of Rome.*

600
City sewer (large scale)
Cloaca Maxima in Rome.
Device-struck coin (electrum alloy)
Ionia, West Turkey. The inscription on the coin reads: 'I am the signet of Phanes'.
Electrical properties (of rubbed amber)
Thales of Miletus, *Greek-Ionian philosopher.*

592
Anchor fluke
Anacharsis, *witty Scythian prince, who travelled widely in search of knowledge.*

570
Geographical/celestial charts
Anaximander of Miletus, *Greek-Ionian philosopher. Also credited with the invention of the gnomon.*
The Hanging Gardens of Babylon
Nebuchadnezzar, *King of Babylon.*

552
Celestial and terrestrial spheres
Anaximander of Miletus.

550
Measurement dial
Anaximander of Miletus.
Silver and gold coinage (inscribed)
Croesus, *King of Lydia, Eastern Mediterranean. Also erroneously attributed to Pheidon, King of Argos and Midas, King of Phrygia.*

540
Harmonic strings
Pythagoras, *wandering Greek philosopher, after hearing four blacksmiths working with hammers in harmony whose weights he found to be six, eight, nine and twelve.*

530
Library
Pisistratus, *'tyrant' of Athens.*

520
Tube structure of the ear
Alcmaeon, *Greek physician and philosopher of Croton. Later known as the Eustachian tubes.*

505
Democracy (practical system)
Cleisthenes, *statesman of Athens, Greece.*
Observatory
erected on top of the temple of Belus at Babylon, possibly used by Naburiannu, the astronomer.

500
Telegraphy (vocal)
King Darius I, *surnamed Hystaspis, of the Persians. He used shouting from hilltops to convey his orders. Mountain horns, trumpets and drums were also used to communicate across distances. So were beacon fires, smoke signals and reflecting mirrors.*

480
Floating bridge(double line of boats)
Xerxes I, *King of Persia. Enabled his army to cross the Hellespont in seven days and nights.*

477
Mnemonics
Simonides of Kos, *called The Younger, a Greek lyric poet working in Athens. He won some 56 poetical contests using his system.*

460
Medical diagnosis
Hippocrates of Kos, *physician.*

447
Chryselephantine statues
Phideas, *Greek sculptor. His statues of Athena, Parthenos and Zeus ranked among the Seven Wonders of the World.*

420
Regal portrait on a coin
Kharai, *King of Lycia.*

Silver tetradrachma coin, used in Athens, 430 BC

400
Automaton flying dove
Archytas of Tarentum, *Greek general and mathematician.*
Gastraphetes ('stomach-bow')
Greek military engineers in Sicily.
Ionian-Greek alphabet
24 letters.
Kite
China.
'Knives and Hoes' (copper coinage)
China.
Pulley
Archytas of Tarentum.
True arch (architectural device)
Palera, Greece.

350
Mausoleum
Mausolus, *King of Caria and husband of Artemisia, who erected a magnificent monument to his memory at Halicarnassus.*
Three conic sections
Menaechmus, *Greek mathematician.*

334
Private library
Aristotle, *Greek philosopher, scientist and physician. He made use of this to educate Alexander, the son of Philip of Macedon – and then at his school, the 'Lyceum' in Athens.*

320
Euclid's Elements
Collated by Euclid, *Greek mathematician.*

312
Aqueduct (long-distance)
Appius Claudius Caecus. *He also constructed the Appian Way.*

300
Aryan alphabet
India.
Canals and rivers (controlled by a series of gates)
China.

Crossbow (individual weapon)
China.
Disc coins (with holes)
China.

300–200
Trip-hammer mill
China.
Ultra-straight road network
Roman engineers throughout Italy.

285
Lighthouse
'The Pharos', near Alexandria, Egypt, was 550 ft high and visible for 42 miles.

280
The Colossus of Rhodes
Chares, *sculptor and engineer. This gigantic figure of Apollo, 700 cubits high, its legs astride the harbour for ships to pass in and out, was destroyed by an earthquake in 224 BC.*

260
**Archimedean screw
(mechanical lifting device)**
Archimedes, *son of the astronomer Pheidias, whilst studying in Alexandria.*

Archimedean screw

Bibliographic system
Callimachus, *Greek poet and scholar. His 'Pinakes' (tablets), running to 120 books, were an elaborate catalogue of the Great Library at Alexandria.*

250
Hydraulis (water-powered organ)
Ktesibios, or Ctesibius, *physicist and former barber of Alexandria.*
Fire-engine
Ktesibios.
Quinqueremes (war galleys)
several Eastern Mediterranean city-states.
The thirty systematic dynastic divisions
Manetho, *Egyptian priest in Alexandria. Later proved incorrect.*
Toothed wheels
attributed to Ktesibios.

240
Lithotomy ('cutting for the stone')
Ammonius, *Greek surgeon.*

225
Armillery sphere
Eratosthenes, *Greek astronomer, geographer and mathematician working at the Library in Alexandria.*

221
Gunpowder
Alchemists and engineers *in Chin Dynasty, China.*

220
Ellipse and hyperbola
Apollonius of Perga, *Greek mathematician.*
Hydrostatics (fundamental law)
Archimedes of Syracuse *on the island of Sicily to check the extent to which the crown of his distant kinsman, King Hieron II, was pure gold.*
World map
Eratosthenes, *superintendent of Ptolemy Euergetes's library at Alexandria. He also invented the 'Cribrum Arithmeticum'.*

215
Leverage (large-scale)
Archimedes of Syracuse. *For*

defensive war machines to ward off the Roman invasion as led by Marcellus. Catapults and grappling irons, based on lever devices, inspired fear and awe among the Romans for almost 3 years.

214
Planetarium (table size)
Archimedes of Syracuse.

213
Solar energy mirrors
Archimedes of Syracuse. *A barrage of reflectors along the harbour walls which not only dazzled Marcellus's approaching fleet but set fire to the sails and timber work.*

206
Reconnaissance kite
General Han-Sin, *to ascertain the distance between his camp and the palace Wei-Yang-Kong, in China.*

200
Glass-blowing iron
Babylonian glassworkers.

198
Vellum books
Attalus I, *King of Pergamum, for his great library which, under the direction of Crates of Mallus, grew to 200,000 volumes, many of them parchment.*

196
The Rosetta Stone
bilingual text in three scripts: hieroglyphs, demotic and Greek.

193
Warehouse (multiple barrel-vault design)
Porticus Aemilia (the Aemilia gate).

190
Hexameter (verse line of six metrical feet)
Quintus Ennius, *Roman poet.*

Pyrphoros (flame-thrower)
Pausistratus, *a Rhodian admiral.*

142
Stone bridge
Roman engineers, *across the River Tiber, Rome.*

130
Astrolabe
Hipparchus, *Greek astronomer of Rhodes.*
Trigonometry
Hipparchus.

121
Concrete building
The Temple of Concord, Rome.

100
Draw-loom for figured weaves
China.
Sails and rigging (mat-and-batten principle)
China.

Silk-reeling machinery
China.
Wheel-bearings
four-wheeled wagon found at Dejbjerg, Jutland.

85
Seed-drill plough (with hopper)
China.

50
Caesarian Cipher
Gaius Julius Caesar. *Also used by the Emperor Augustus.*

46
Julian Calendar
Sosigenes of Alexandria, *astronomer and mathematician, employed by Julius Caesar to devise a more accurate calendar.*

40
Rotary winnowing machine (with crank handle)
China.

17
Lithotomy (small apparatus)
'Celsus', alias Aulus Cornelius, *Latin physician.*

20
Trip hammer mill (water-powered)
China.

31
Metallurgical blowing engine (water-powered)
Tu Shih, *Prefect of Nanyang, China.*

60
Miniature reading lens
the Emperor Nero, *who used the large transparent gemstone in one of his rings.*

83
Lodestone spoon rotating bronze plate (magnetism)
China.

c. 100
Whirling Aeoliphile (steam-powered)
Hero, *Greek mathematician of Alexandria. Among his other inventions were an olive oil beam press, a grape screw press, a screw-cutting machine, a gastraphetes (crossbow), water- and wind-powered musical organs, automatically opening doors and even a Holy Water slot machine. 80 such gadgets were described in his*

Hero's whirling Aeoliphile

tome, 'Pneumatica'.
Earthen lamp
Epictetus, *former Roman slave and Stoic philosopher of Nikopolis in Epirus. The invention was sold after his death for 3000 drachmas.*

105
Paper-making (mulberry-based)
T'sai Lun, *eunuch director of the Imperial workshops of the Chinese Han Court and later deified as the god of paper-makers.*

124

124
Pantheon (large circular temple)
Architects and engineers of the Roman Emperor Hadrian.

132
Earthquake alarm
Chang Heng, *Chinese Imperial astronomer, mathematician, cartographer and poet. One of the wonders of the Imperial Observatory, it was 8 ft high and incorporated a free-swinging pendulum.*

150
Map of the world (latitude and longitude grid system)
Claudius Ptolemaeus, *Greek astronomer and geographer.*

170
Edge-runner mill
China.

180
Gimbals
China.
Rotary fan for ventilation
China.
Zoetrope (lamp cover revolved by ascending hot air)
China.

189
Square-pallet chain-pump
China.

200
Artery (as blood carrier)
Galen of Pergamum, *overturning the belief that they were merely air tubes.*

224–241
Circular-plan city
King Ardashir I *at (present-day) Firuzabad, south-west Iran.*

231
Wheelbarrow
China.

239
Jewish Year 4000.

320
Helicopter top (spun by pulling a cord)
China.

340
Wagon-mill (grinding during travel)
China.

366
Prayer beads
attributed to Saint Augustine of Hippo.

370
De Rebus Bellicis
anonymous author. *Containing working drawings for defence machines against the Vandals by Valentinian I, Roman Emperor, and his brother, Valens. Includes an ox-powered paddleboat.*

372
Public hospital
Saint Basil, *Bishop of Caesarea in Cappadocia.*

385
Jewish calendar
Rabbi Hillel II.

400
Edge-runner mill (water-powered drive)
China.
Ship (watertight compartments)
China.

410
Commemorative metal crucifix
On a rocky mount of the open court between the basilica and rotunda of the Holy Sepulchre in Jerusalem. Seven years later, Christian Roman Emperor Theodosius II encased the crucifix with

gold and precious jewels. Soon after, crucifixes started to appear in and on churches.

450
Whisky
attributed to Saint Patrick, *although 'Aqua Vitae' had been brewed by the Romans 500 years before.*

512
Arabic alphabet: 28 letters.

520
Decimal system
Aryabhata and Varamihara, *Indian mathematicians.*

532
Anno Domini (dating system)
Dionysius Exiguus, *Scythian abbot of a monastery near Rome. Not generally employed for several centuries.*
Single-shell dome
Anthemios of Trales, *architect for Saint Sophia Cathedral.*

535
Stained glass windows
Saint Sophia Cathedral.

550
The Great Arch of Ctesiphon

552
Sailing carriage (first high land speeds)
China.

555
Watermill (outside China)
Belisarius, *general of the Emperor Justinian, while besieged in Rome by the Goths.*

580
Iron-chain suspension bridge
China.

610
Segmental arch bridge
An-chi bridge at Chao-hsien, China. Built by the Sui Emperor Wen-ti *to span the Grand Canal, dug to join the Yang-tze and Yellow River systems.*

640
Ching-Chow Astronomical Observatory
Korea, during the reign of Queen Sondok of Silla.

644
Windmill
Persia, for grinding corn.

672
'Greek Fire'
Kallinikos, *an architect from Heliopolis, Syria, to help the Byzantines destroy the Arab Saracen ships. 30,000 men reported killed by this weapon.*

Byzantines use 'Greek Fire' against Saracens

700
Porcelain (original)
T'ang Dynasty, China.
Ship (stern-post rudder)
China.

770
Block-printed Buddhist charm.

800
Carolingian minuscule script
various scholars under the direction of Alcuin, *Anglo-Saxon abbot of Tours and adviser of the Frankish emperor, Charlemagne.*

808
Bank (money repository)
Jewish merchants in Lombardy.

825
True lock-gates and chambers
China.

862
Slavonic alphabet
Saint Cyril, *alias Constantine the
Philosopher, to convert the Moravians
to Christianity. 42 letters.*

868
Printed book
*The Diamond Sutra, found in
Buddhist caves at Tun-Huang, China.*

879
Gorgonzola cheese
farmers in the Po Valley, *Italy.*

886
**Wax candle 24-hour
measurement system**
Alfred the Great, *King of the West
Saxons and overlord of England. Six
candles, burning at 3 inches per hour,
were used each full day.*

890
Nailed-on horseshoes
Siberia, Byzantium and Germany.

900
**Concave, curved iron
mouldboard (plough)**
China.

910
Paper money
China.

970
Central Library of Cordoba
Abd-al-Hakam II, *Caliph of
Cordoba, Spain, in his Alcazar (royal
palace). Over 400,000 titles
catalogued in 44 volumes.*

983
Canal locking system
Chiao Wei-Yo, *China.*

990
Giant sextant (radius 55 ft)
Hamid Ibn al-Khidr al-Khujandi,
Arab astronomer.

995
Giant quadrant (22 ft radius)
Abdul-Wafa Albuzjani, *Persian
mathematician.*

1000
Venetian glassmaking system
*first made in the city and then, due to
the danger of fire, transferred to
nearby Murano island.*
Coffee
reported by Avicenna, *Arabian
philosopher and physician.*

1024
Sonnet (poetical form)
Guido d'Arezzo, or Aretino,
*Benedictine monk and musical theorist,
at Pomposa, near Ferrara, Italy. It
was later standardised by* Francesco
Petrarch *(born Arezzo, 1304).*

1026
Musical notation
Guido d'Arezzo.

1049
Printing (clay characters)
Pi Sheng, China.

1090
**Silk-flyer (for twisting and
doubling)**
China.

1092
Astronomical clock
Su Sung, *after five years' work at the
Imperial Observatory at K'ai Feng. A
cross between the clepsydra and the
spring-driven clock because its time-
keeping ability could be adjusted by*

Frankish warrior, c. 1100

weights. It struck a gong to indicate the passing hours, by a system of bamboo 'revolving and snapping springs'.

1100
Rockets and fire-lances
China.

1115
Lodestone-magnetic compass for navigation
China.

1150
Paper-mill
Xativa, Spain.

1155
Oldest printed map of China.

1200
Automaton dancing men and walking peacock
Al-Jazari, *Arabian scholar.*
Ormulum
Ormin, *an Augustinian monk, versifier and spelling reformer, living in the east of England. It was a metrical version of the Gospels and the*

Acts of the Apostles.
Rhythmic musical notation
Franco of Cologne *in his 'Art of Measurable Music'.*

1202
Rosary (prayer beads)
attributed to Domingo de Guzmán, (Saint Dominic).

1220
Practica Geometrica
Leonardo of Pisa, alias Fibonacci.

1221
'Iron fire bombs'
the Kin warriors of China.

1239
Jewish Year 5000.

1241
Hanging, drawing and quartering
first inflicted in England on William Marise, a pirate and nobleman's son.

1245
Mechanical eagle
Villard de Honnecourt. *Turned its head towards the deacon when he read the Gospel.*
Overbalancing wheel (perpetual motion machine)
Villard de Honnecourt, *French architect. One of 325 working drawings in his sketchbook.*

1250
Magnifying glass
Roger Bacon, *philosopher and scientist, working in Oxford, England. He also designed a lighter-than-air machine, an automatic ship and an automatic carriage.*

1253
Alphonsine Tables
King Alphonso X of Castile, *who spent over 400,000 crowns collecting tables composed by Spanish and Arab*

astronomers. *These tables were to be frequently consulted for the next three centuries.*

1259
Bamboo-gun
('fire-spurting lance')
China.

1260
Biblia Pauperum
(Bible for the poor)
Giovanni di Fidanza, *Italian theologian, General of the Franciscans. Later beatified as Saint Bonaventura.*
Gun/cannon
Konstantin Anklitzen (alias 'Berthold Schwarz'). *? Franciscan, Augustinian or Dominican monk;? of Freiburg, Goslar, Ghent, Mainz, Metz, Cologne, Brunswick and/or Prague – not to mention Danish, Greek, Negro and/or Welsh origins. Other dates given for his/her invention 1313, 1320, 1354, 1359 and/or 1380. Details rather hazy!*

1269
Compass
Petrus Peregrinus de Maricourt. *With 360° card graduation and movable sights for taking bearings.*

1276
Giant astronomical gnomon (40 ft)
Kuo Shou-Ching, *Chinese astronomer. Set in the Tower of Chuo Kung to measure the sun's solstitial shadow.*

1280
Belt-driven spinning wheel
Hans Speyer, *German engineer.*

1285
Spectacles
(convex lens for myopia)
Alessandro de Spina, *Dominican monk at the Convent of Santa Katerina di Pisa, based on the work of Salvino degli Armati, Florentine physician.*

1286
Watermark
(paper identification)
a paper mill in Fabriano, Italy.

1287
Nitric acid (aqua fortis)
Raymond Lully, *Spanish theologian and philosopher.*

1288
Cocchio
(Italian travelling wagon)

1300
Hall-marking
'Gardiens of the Craft', the London Guild of Goldsmiths. *Items of gold and silver struck with a leopard's head, the mark of King Edward I – later a lion's head.*

1305
Acre (standardised)
King Edward I of England.

1310
Blowing-engines for furnaces and forges (crank-handle drive)
China.
Silk machinery (water-powered)
China.

1311
History of the world
Rashid-eddin *of Persia's 'Jami'u't Tawarikh'.*

1313
Indulgences for public sale
Pope Clement V of Avignon.

1326
Iron-bullet-firing metal cannon
Bombards, *as developed by* Rinaldo di Villamagna, *Florence, Italy.*

1330
Caricatures
Buffalmacco Buonamico de
Cristofano, *painter of the early
Florentine School, with a great
reputation as a practical joker. He
drew comic figures and put labels to
their mouths with sentences.*

1335
Automatic striking clock
*installed in the belfry of the Palace
Chapel of the Visconti, Milan, Italy.*
Da Vigevano's Treatise
Guido de Vigevano, *Italian physician
working in France. Contained
drawings of men working the first
known crankshafts inside a road
vehicle, a paddle-wheeled submarine, a
siege tower machine, etc.*

1339
Ship (gun-carrying)
*the 'Christopher', an English cog,
armed with three iron guns and a hand
gun.*

1343
Artillery
the Moors of Algeciras, *Spain. Three
years later King Edward III of
England used four cannons in the
Battle of Crécy.*

1360
**Mechanical clock
(iron movement)**
Henri de Vick *of Württemberg for
King Charles V of France at his royal
palace in Paris.*

1364
Astronomical clock
Giovanni de' Dondi, *astronomer and
physician, for the library of the Duke
of Pavia in Visconti Castle. Showed
the time, the phases of the moon, the
signs of the zodiac and other
astronomical features.*

1367
London's first striking clock.

1381
Padlock
inventor unknown.

1383
Wooden toy cannon
*presented to the child King Charles VI
of France.*

1388
Side-saddles for ladies
introduced by Anne of Bohemia, *wife
of King Richard II of England.*

1392
**Type foundry
(movable bronze characters)**
set up by T"ai Tsung, *King of Korea.*

1410
Wire
Rodolph of Nuremberg, *who devised
the process for drawing out the wire.*

1411
Pasini cipher
Luigi Pasini *of Venice.*

1413
Toy dolls with movable arms
played with by children in Nuremberg

1420
Wagon-fort
Jan Ziska, *leader of the Protestant
Hussites of Bohemia. It was drawn by
teams of horses and armoured with
sheets of iron. Its crew fired crossbows
and guns through loopholes pierced in
the sides. With only 25,000 men and
his wagon-forts, Ziska crushed the
invading force of 200,000 men.*

1421
**Barge with special hoisting gear
(for transportation of building
marble)**
Filippo Brunelleschi, *architect and*

1425

Cocking a cross-bow, 1420

engineer. *Special privilege granted to him by the Signoria of Florence.*

1425
Giant sextant (180 ft high)
Ulugh-Beg, *Persian prince and astronomer, for his observatory at Samarqand.*

1432
Dome (negative side-thrust)
Filippo Brunelleschi, *architect and engineer, for the Duomo of his native Florence.*

1437
Perspective views (pinhole demonstration)
Leon Battista Alberti, *Italian architect.*

1440
Playing cards (the four suits)
Provence, France. But known as far back as 950 AD in China, supposedly developed by harem women to stave off boredom.

1444
Waterless mills
exclusive privilege granted to Antonius Marini de Francia *to construct these by the Venetian Senate.*

1451
Printing press
Johannes Gansfleisch zur laden zum Gutenberg *in Mainz, Germany, with the financial backing of goldsmith Johann Fust and assistance from Peter Schoeffer. Following the printing of an old German poem, Gutenberg threw his energies into the printing of a 1282-page Bible in Latin – an enterprise which bankrupted him out of his workshop.*
Travelling clock
Jean Lieburc, *Paris, with its copper casing by Jean Moulinet.*

1452
Baptistry doors (bronze)
Lorenzo Ghiberti, *sculptor and bronze-caster in Florence.*

1455
Cast-iron pipes
to carry water at the Castle of Dillenburgh, Germany.

1458
Bereguardo Canal
Bertola da Novate, *Italian civil hydraulic engineer. 12 miles long with a complete set of canal locks.*

1464
De Triangulis Omnimodis
Regiomontanus, alias Johannes Müller, *mathematician and astronomer of Nuremberg.*

1470
Jenson typeface
Nicholas Jenson, *French printer and
type-founder in Venice.*
Page numbers in a printed book
Werner Rolewinck's '*Sermo ad
Populam*', *published by Hoernen of
Cologne.*

1472
Printed music
'*Gradual*' *by unknown Constanz
printer.*

1474
Automaton eagle and iron fly
Regiomontanus, alias Johannes
Müller, *German mathematician and
astronomer of Nuremberg.*
Patent Law
The Venetian Senate, *to protect
inventors.*

1475
Hourglass set
*Nuremberg. Four different sizes to run
15, 30, 45 and 60 minutes.*
Lithotomy (high apparatus)
Germain Colot, *French physician, on
a criminal in Paris.*

1476
Italic typeface
Francesco Griffo, *typecutter of
Venice; used for a series of classics
published by Aldus Manutius.*

1477
Printed map
Bologna, Italy.
Printed poster
William Caxton, *London printer – to
advertise the thermal cures at
Salisbury.*

1478
Set of printed maps
Conrad Sweynheym's *27 maps in
Ptolemy's 'Cosmographia', printed by
Arnold Buckinck, Rome.*
Spectacle manufacture
Nuremberg.

1481
Wallpaper
Jean Bourdichon, *painter and
illuminator, who painted 50 great rolls
of paper in blue with the inscription
'Misericordias Domini in aeternum
cantabo' and three angels about 3 ft
high to hold the rolls in their hands.
Bourdichon was paid 24 livres for his
work.*

1484
Mine drainage system
Blasius Dalmaticus *from Ragusa.*
'*Patent rights' were granted him by the
Elector Ernst of Saxony.*

1487
Carillon (bells)
First heard in Alost, *East Flanders.*

1488
Multiple crossbow gun
Leonardo da Vinci, *Italian painter,
architect and engineer, in the employ of
Lodovico Sforza, 'Il Moro' of Milan.
For the previous six years da Vinci, in
his thirties, had been keeping detailed*

Leonardo da Vinci

1489

notebooks, which in special mirror-handwriting show inventions for a longbow shield, a human flying machine, a helicopter, a parachute, a diving bell and suit, articulated chains, giant crossbow, circular armoured vehicle, scythe-wheeled chariots, etc. None of these appear to have been built.

1489
Plus (+) and minus (−) signs
Johann Widmann's *'Mercantile Arithmetic', published in Leipzig.*

1490
Antimony
Basil Valentine, alias Thölde, *German salt manufacturer and chemist.*

1492
Behaim's terrestrial globe
Martin Behaim, *navigator and geographer, with* B Glockenthon, *painter – both at Nuremberg.*

1498
Toothbrush
China.

c. 1500
Bassoon
Afranio, *canon of Ferrara.*

1500
Caesarian section
Jacob Nufer, *sow-gelder of Sigershauffen, Switzerland, on his own wife, using his sow-gelding instruments.*

1501
Hymn-book
Severin of Prague *for the Hussites of Bohemia. Contained 89 hymns in the Czech language.*

1502
Barrel organ
engineers, carpenters and musicians working for Archbishop L. von

Keutschah *of Salzburg, nicknamed 'the 'Salzburg Bull'.*

1510
'The Nuremberg Egg' (round pocket watch)
Peter Henlein, *clockmaker of Nuremberg, after ten years' work, including his invention of a steel mainspring.*

1512
Washing apparatus (for ore)
Sigmund von Maltitz.

1515
Galleas
King Henry VIII of England. *'The Great Galley', 800 tons, with 120 oars manned by 160 oarsmen and 207 guns.*

1518
Fire-engine
Anthony Blatner, *goldsmith for the city of Augsburg.*

1520
Rifle-barrel (spirally-grooved)
August Kotter, *Germany.*

1525
Dental forceps (for extraction)
'Thomaseus'.
Harquebus
Marquis of Pescara, *Spain. The first portable fire-arm: a shotgun having a matchlock operated by a trigger.*

1528
Hodometer
Jean Fernel, *instrument maker, France.*

1529
Bismuth (metal)
'Agricola' alias Georg Bauer, *physician and mineralogist in Joachimstal, Germany.*
Musical bars
'Agricola' in 'Musica Instrumentalis'.

Wandering organists, 1502

*Mechanised power –
ball and chain mining pump, 1529*

1530
Cristallo glass
Venice.

1531
Diving bell
Guglielmo de Lorena, *Italian
engineer, in an attempt to raise
Caligula's pleasure galleys from the
bottom of Lake Nemi.*

1537
'The New Science'
Niccolo Fontana, *nicknamed
'Tartaglia', Professor of Mathematics
at Venice University. Demonstrated,
among other things, how cannon balls
follow a curved path rather than
Aristotelian straight lines.*
**Trucked muzzle-loading gun
carriage**
*discovered on board the 'Mary Rose,'
warship of King Henry VIII of
England.*

1538
Optic nerves
Costanzo Varolio, *anatomist and
physician of Bologna, Italy.*

1540

ANNO·ÆTATIS·
68

Ambrose Paré

1540
Artificial limbs
Ambrose Paré, *French military barber surgeon, the first to replace amputated limbs with sophisticated mechanical substitutes, using hinges, cog wheels, levers and armatures.*
Fusée (watch-spring device)
Jacob Zech *of Prague.*
Meridionel instrument
'Nicolaus Copernicus', *the Latinised name of* Mikolaj Kopernik, *canon of Frombork, Poland. He used it while evaluating his Heliocentric theory.*
Pistol
Camillo Vettelli, *gunsmith of Pistoia, Italy.*

1541
Zinc
'Paracelsus', alias Theophrastus Bombastus *of Hohenheim.*

1543
Detailed human anatomy book
Andreas Vesalius, *Belgian anatomist, assisted by draughtsman* Jan Stephan

van Calcar, *a pupil of the painter Titian.*

1545
Conservatory
Daniel Barbaro *for the Botanical Gardens at Pavia, Italy. A viridarium, heated in winter by a brazier or open hearth.*
Equation of the third degree
Geronimo Cardano, *Professor of Mathematics and Medicine at Padua, Pavia and Bologna.*

1549
Venetian system for making crystal glass
patent protection granted to Jehan de Lame *of Cremona, dealer, residing in Antwerp, to introduce the technique into Brabant.*

1550
Ligature
Ambose Paré, *French military barber surgeon, to replace the blind and brutal use of red-hot cautery.*
Screwdriver
By gunsmiths and armourers, to adjust their gun mechanisms.
Spanner
First designs appeared during this century to deal with the newly developed system of nuts and bolts.

1556
Sealing wax (including shellac).

1557
Enamels
Bernard Palissy, *after sixteen years' research at Saintes, France.*
Equals sign (=)
Robert Record, *Fellow of All Souls College, Oxford, England, in his algebra text 'The Whetstone of Witte'.*

1558
Dolls' house
made by craftsmen for the daughter of the Duke Albrecht of Saxony.

1560
Camera obscura (portable)
Giovanni Battista della Porta,
Neapolitan physicist and philosopher.
Wallpaper (block-printed)
printers *in Delft, Holland and*
London, England where there was a
Guild of Paperhangers.

1561
Dredger
Pieter Breughel *for Brussels*
Municipality. Used to excavate the
Rupel-Scheldt canal.

1563
Wire-manufacturing mill
Nuremberg.

1565
Graphite pencils
Konrad Gesner, *the year after the*
discovery of the Borrowdale deposits in
Cumberland, England by German
miners prospecting for copper.

1567
Silk draw looms
first British use in Norwich.

1568
Mercator Projection (map
treating the globe as a cylinder
rolled out flat)
Gerhard Kremer, *Flemish*
mathematician, geographer and map-
maker.

1569
Screw-cutting and ornamental-
turning lathes
Jaques Besson, *engineer in the Court*
of King Charles IX of France.

1570
Log line (for navigation).
Modern atlas
Abraham Ortelius *of Antwerp.*
'Theatrum Orbis Terrarum' contained
70 maps on 53 plates.

1571
Theodolite
Leonard Digges, *mathematician,*
England.

1575
Paper tape-recorder (compass
bearings)
Thomas Ruckert, *England.*

1576
Magnetic needle dip
Robert Norman, *instrument-maker of*
London.
'Uraniborg' Observatory
Tycho Brahe, *Danish astronomer.*
Sited at Hven Island, off the Swedish
coast.

1578
Prefabricated building
Nonesuch House, *brought from*
Holland in sections and re-erected on
London Bridge.

1581
Pendulum motion
Galileo Galilei, *who timed the*
oscillations of the great lamps of Pisa
Cathedral, Italy, swinging from the
roof during an earthquake tremor –
using his pulse beats.

1582
Calendar reform
Christoph Clavius, *German Jesuit*
mathematician, at the request of Pope
Gregory XIII.
Pumping station (water supply)
Morris's London Bridge Water
Works.

1585
Fire-ship (explosion vessel)
to destroy a bridge of boats at the siege
of Antwerp. Three years later Lord
Howard of Effingham *used them for*
the engagement with the Spanish
Armada.

Time bombs
Clockwork-activated floating mines used by the Dutch at the afore-mentioned siege of Antwerp.

1586
Giant cannon (bombard)
'The Tsar' of Moscow, *with a 36-inch bore.*

1588
'Reading machine' (vertically revolving bookcase)
Agostino Ramelli, *Italian engineer working in France. One of 195 machines described in a book he published that year in Paris.*

1589
Hosiery-knitting machine
Reverend William Lee *of Woodborough, Nottingham, England. He was refused patronage to develop it, by both Queen Elizabeth I and the London city merchants.*
Stamp mill for hemp (powered by wind, horse, or water)
Elbert de Veer, *burgher of the City of Amsterdam.*
Water closet
Sir John Harington, *godson to Queen Elizabeth I of England and amateur poet. He had been banished from the Court for circulating a risqué story. During exile he built a house at Kelston, Bath and there installed the first flushing lavatory. He next supplied one to his royal godmother at Richmond Palace apparently as a ploy for a literary exercise, for he described his device in a work called 'The Metamorphosis of Ajax – a Cloacinian Satire'. A 'jacks' was a privy. Thus rivalry to the chamber pot was set in motion.*

1590
Compound microscope
Hans Janssen *and his son* Zacharias, *spectacle-makers of Middleburg, Holland. With the use of concave and convex lenses it was soon improved by Galileo Galilei.*

1592
Armoured ship
Korean Admiral Yi-sun-Sin *for use against the Japanese. With this ship, iron-plated turtle-back, covered with rows of spikes and fitted out with both a ram and archery ports for fire arrows, Yi-sun-Sin virtually wiped out the Japanese fleet in a matter of days.*
Timber-sawing machine (windmill-powered)
Cornelis Corneliszoon *of Uitgeest, Holland.*
Windmills for shammy dressing
Henrick van Zanthen *and* Franz Lambrechtz *from Bois le Duc, residing in Amsterdam.*

Windmill-power

1593
Transportable horse-driven flour mill
Jooris Fries van Embden *and* Merten Jansz *of Dortmund in Westphalia.*

1594
The Back-staff
Captain John Davis, *English navigator and Arctic explorer. Became known as 'Davis's Quadrant'.*

1595
'Flight ship' (swift flat-bottomed yacht)
Pieter J Livorn, *boatbuilder, Hoorn, Netherlands.*
Map atlas
'Gerardus Mercator', alias Gerhard Kremer, *Flemish mathematician, geographer and map-maker.*

1596
'Oorgat' (slot in bridges to allow ships to pass under bridge with mast up)
Hendrick Jacobzoon Staes, *carpenter, Amsterdam.*
Recording map-maker (hodometer-compass)
Christoph Schissler, *instrument-maker of Augsburg.*

1597
Graphometer
Philip Danfrie, *France, chief engraver at the Paris Mint.*

1598
Eternally-running clockwork
Cornelis Jacobzoon Drebbel *of Alkmaar, Netherlands.*

1599
Flageolet (musical instrument)
Juvigny, *instrument-maker of Paris, as a variant of the recorder.*
Silk-knitting machine
Reverend William Lee *of England. Following Lee's obscure death in the Low Countries in 1610, his brother returned to England, and with patronage from a Nottingham merchant, founded the first automatic frame-knitting factory in the North Midlands.*

Stevin's sailing chariot

1600
Sailing chariot
Simon Stevin, *mathematician for Stadtholder Prince Maurice of Nassau. The two-masted chariot is credited with a speed of 21 mph on a two-hour journey from Scheveningen to Petten in Holland, with 28 passengers and crew on board.*

1601
Improved printing press
William Blaen *of Amsterdam.*

1602
Measuring compass
Jost Bing *of Hosso.*

1603
Pantograph
Christoph Scheiner, *German Jesuit, for copying, reducing or enlarging plans. An instrument of pivoted levers.*

1605
Multiple-ribbon weaving mill
Willem Dircxz van Sonnevelt *of Leyden, the Netherlands.*

1606
Surveying chain
Edmund Gunter, *English*

mathematician and astronomer of
Christ Church, Oxford, England.

1608

Hanging compass
William Barlowe, English divine and
natural philosopher.
Telescope
Hans Lippershey, spectacle-maker of
Middleburg, Holland. Demonstrated
his 'kjiker' (looker) before the
Netherlands States General and was
subsequently paid 900 florins for the
invention.

1609

Astronomical telescope
Galileo Galilei, using lenses fashioned
by the glassworkers of Murano,
Venice. Inspired after hearing of
Lippershey's 'looker', Galileo used his
instrument to discover the mountains of
the moon, and the satellites of Jupiter.
Newspaper (regularly printed)
simultaneously by Julius Adolph von
Söhne, Keeper of the Royal Press,
Saxony, and Johann Carolus of
Strasbourg. Von Söhne's 'Aviso'
lasted until 1624 whilst Carolus's
'Relation' had stopped production two
years before.
**Thermostatic incubator (for
hatching eggs)**
Cornelis Jacobzoon Drebbel, Dutch
physicist working in London.

c. 1610

Blood circulation
William Harvey MD, Pavia and
Cambridge, physician to St
Bartholomew's Hospital, London. He
published his findings in 1628.

1610

Flintlock gun
Marin Le Bourgeois, France.
'Ludolph's Number'
Ludolph van Ceulen, mathematician
and Professor of Fortification at
Leyden, who discovered it whilst in
search of the value of π (pi).

Saturn observed by telescope
Galileo, using a magnification of 32 on
his largest instrument.

1611

Coke
Simon Sturtevant, ironmaker,
England.
**Microscope (double convex
compound)**
Johann Kepler, Imperial
mathematician, Prague, Bohemia;
practical instrument built in 1628 by
Christopher Scheiner.
Rainbow (theory of)
Johann Kepler.
**Telescope (double convex and
darkened glass)**
Christoph Scheiner, German Jesuit
at Ingolstadt. Darkened glass enabled
him to observe sunspots.

1612

**Circumferentor ('Holland
Circle')**
Jan Dou, Amsterdam, Holland.

1613

Astrolabe
David Fabricius, astronomer and
pastor at Resterhaave and Osteel in
East Friesland.

1614

Logarithms
John Napier, mathematician, inventor
and astrologist of Merchiston,
Scotland. He also invented a
calculating apparatus nicknamed
'Napier's Bones'. The decimal base for
logarithms was invented four years
later by Henry Briggs of Oxford,
England.

1615

Oval-turning lathe
Salomon de Caux, engineer to Henry
Prince of Wales, later King Charles I
of England.
Solar motor
Salomon de Caux, for the

entertainment of royalty. Used to power a fountain.
Surveying by instrumental triangulation
Willebrord Snell van Roigen, *Professor of Mathematics at the University of Leyden, Holland. He measured an arc of meridian.*

1616
Medical thermometer
Santorio Santorio, *Professor of Theoretical Medicine at Pavia, Italy. It was an adaptation of the air thermometer of his friend, Galileo Galilei. Santorio also invented a pulsimeter.*

1617
Coin-milling engine
Jean Balancier *of France.*
Map manufacture (British towns)
Aaron Rapburne *and* Roger Burges. *British Patent Number 1.*
Oil (to keep armour and arms from rust or canker)
John Casper Wolfen *of King James I of England's Privie Chamber, with Johannes Müller from Germany.*

1618
Ploughing machine
David Ramsay *and* Thomas Wildgoose, *England.*

1620
Double bassoon
Hans Schreiber *of Berlin.*
Merry-go-round
at a fair at Philippopolis, Turkey.
Submarine boat
Cornelis Jacobzoon Drebbel, *Dutch physicist working in London. This was a wooden boat covered with leather and smeared with tallow to render it watertight. It was powered by twelve oarsmen, breathing air from the hollow mast. Demonstrated to King James I in 1624.*

1621
Slide rule (rectilinear)
William Oughtred, *Rector of Albury, Surrey, England. He also invented the trigonometric abbreviations, 'sin' and 'cos'.*

1622
Lacteals
Gaspare Asellio, *Professor of Anatomy and Surgery at Pavia, Italy.*

1623
Drainage or flour mill (driven without wind or manual or animal power)
Lieutenant Laurens Garlyck, *the Netherlands.*
Practical mechanical calculating machine
Professor Wilhelm Schickard *at Tübingen University, Germany.*

1625
Hackney coaches (for hire)
four set up in London by Captain Bailey, *retired sea-captain.*

1626
'Leather' gun
Gustavus Adolphus *of Sweden – complete with internal copper bore.*

1630
Slide rule (circular)
Richard Delamain, *English mathematician, former pupil of William Oughtred.*

1631
Classified advertisements
Théophraste Renaudot, *France, in 'La Gazette'.*
'Fermat's Last Theorem'
Pierre de Fermat, *mathematician and Commissioner of Requests, Toulouse, France.*
Multiplication sign (×)
William Oughtred *of Albury, Surrey, England in his 'Clavis Mathematica'.*

Vernier Auxiliary Scale
Pierre Vernier, *French-born mathematician working for the King of Spain in the Low Countries.*

1632
Automaton lady and cavalier
Philip Hainhofer. *For an art cabinet presented by the town of Augsburg to King Gustavus Adolphus.*
Fire engine ('water-bow')
Thomas Grent.

1634
Coloured-fabric cement
Jerome Lanyer, *textile merchant, England. Called 'Londrindiana'. He also invented 'velvet paper'.*
Combined bow and pike weapon
William Neade, *skilful archer, England.*
Sedan chair
Sir Saunders Duncombe, *who patented and promoted this mode of transport and acquired sole privilege to use, let and hire a number of them for fourteen years.*

1635
Ship salvage engine
Mathewe van Dyck, *Dutch mechanic working in London.*

1636
Blade manufacture (mill-powered)
Benjamin Stone, *swordblade manufacturer, England.*

1638
Micrometer
William Gascoigne, *England (killed at the Battle of Marston Moor in 1644). To measure the dimensions of the sun, the moon and the planets.*

1640
Magic lantern
Athanasius Kirchner, *a German Jesuit.*

Taxi
Nicholas Sauvage, *rue de Saint-Antoine, Paris, France, who had twenty carriages for hire.*

1642
Calculating machine
Blaise Pascal, *nineteen-year-old son of the local president of the Clermont-Ferrand court of the exchequer. To assist his father with his accounts. Patented in 1647, only seventy 'sautoirs' were made, some of them as presents for King Louis XIII of France.*
Mezzotint
Colonel Ludwig von Siegen, *who engraved a portrait of Princess Amelia of Hesse. He disclosed the process to Prince Rupert at Brussels in 1654 and the latter subsequently had craftsmen improve upon it.*

1644
Barometer
Evangelista Torricelli, *Professor to the Florentine Academy, with his colleague* Vincenzo Viviani. *At first known as the 'Torricellian Tube'.*
Monocable ropeway
Adam Wybe, *Dutch engineer, to facilitate the building of the fortifications of Gdansk.*

1647
Bayonet
An unknown armourer at Bayonne, France.
Thoracic duct
Jean Pecquet, *medical student, Montpellier University, France.*
Map of the moon
'Hevelius', alias Johannes Hewelcke *of Danzig, who drew up a catalogue of 1500 stars and observed the planets, the moon and comets from his private observatory.*

Torricellian Tube, later barometer (1644)

Newspaper (daily)
Timotheus Ritzsch *of Leipzig, whose 'Einkommenden Zeitungen' was in production for three months.*
Wheelchair (tricycle)
Johann Haustach *of Nuremberg for the legless Stephen Farfler, who propelled it with hand-cranks.*

1653
Invisible ink recipe
Peter Borel, *although 'sympathetic' or 'secret' inks had been in use since 300 BC.*
Letter-boxes
François Velayer *of Paris, for his 'petite poste'.*

1654
Air vacuum pump
Otto von Guericke, *engineer of Magdeburg. Publicly demonstrated in front of the Imperial Diet of Ratisbon.*

1655
'A Century of the Names and Scantlings of Such Inventions As ... I ... Have Tried and Perfected'
Edward Somerset, *2nd Marquis of Worcester. Written while he was on bail from the Tower of London, it gives a brief account of 100 inventions, including a steam engine and a 14-foot diameter perpetual-motion wheel.*

1648
Hydraulic press
Blaise Pascal, *mathematics and physics student, Rouen, France. He also invented the syringe about this time.*
Hydrochloric acid
Johann Rudolph Glauber, *wandering chemist.*

1649
Automaton coach and horses
Camus *for Louis XIV of France when a child. It had a footman, a page and a lady inside. The horses and figures moved naturally and perfectly.*
Frigate (warship)
English naval architects.

1650
Lymphatic glands
simultaneously by Olavs Rudbek *in Uppsala, Sweden,* Thomas Bartholin *in Denmark and* Jolyffe *in England.*

1657
Pendulum clock
Christiaan Huygens, *Dutch physicist whose timekeeper was built at the Hague by Salomon Coster and installed at Scheveningen. Huygens was the first man to patent the pendulum for clock manufacture.*

1658
Balance spring regulator
Robert Hooke, *engineer and friend of Sir Christopher Wren. Hooke designed a clock mechanism regulated by a balance-spring, whose elasticity caused the balance-wheel to oscillate in equal*

times. A company was formed to manufacture the watch but failed due to mistrust.
Red blood corpuscles
Jan Swammerdam, *Dutch naturalist and pioneer of microscopy in Amsterdam, Holland.*

1659
Cheque (monetary)
Messrs Clayton & Morris, *bankers, law scriveners and estate agents of Cornhill, London. For £400 by Nicholas Vanacher, made payable to Mr Delboe.*

1660
'Berlin' (horse-drawn carriage).
Frictional electrical machine
Otto von Guericke, *engineer of Magdeburg.*
Twin-quill pen
Sir William Petty, *political economist working in Ireland.*

Otto von Guericke's 'Magdeburg Hemispheres'

1661
Methyl alcohol
The Hon. Robert Boyle, *Irish physicist and chemist, working in Oxford, England.*
Postal date-stamp for letters
England. Penny post first set up in London and its suburbs, two years

later, by Robert Murray, *upholsterer.*
Bank notes
Issued by the Bank of Stockholm, Sweden.

1662
'Boyle's Law'
The Hon. Robert Boyle.
Buses (eight-seaters)
Blaise Pascal, *French philosopher and scientist, with his friend and chief financial backer, the* Duc de Roannez. *First scheduled service ran between Porte de Saint-Antoine and Porte de Luxembourg, Paris. Called 'carrosses à cinq sous' (five-cent coaches).*
Catamaran yacht
Sir William Petty, *founder member of the Royal Society of England. The following year, crewed by 30 men, 'Experiment' won the first open yacht race.*
Spirit level (air-bubble system)
Jean de Mechisedech Thévenot, *French researcher and traveller.*

1664
Hygrometer
Francesco Folli, *Italy.*
Lens-grinding lathe
Giuseppe Campani, *engineer of Rome.*
Theatre curtains
Ichimura-za Theatre, *Edo, Japan.*

1665
Blue sugar paper
Charles Hildeyard, *England.*
Syphon wheel barometer
Robert Hooke, *Curator of the Royal Society, Oxford University. Further developed in 1670 by* Sir Samuel Morland, *'master mechanic' to King Charles II.*

1666
'Fluxions'
(principles of integral calculus)
Isaac Newton, *24-year-old mathematics graduate of Trinity College, Cambridge, England.*

**Open-plan city
(piazza-radiating streets)**
Sir Christopher Wren, *architect and
Surveyor-General, following the
Great Fire of London. Never
constructed due to vested interests.*

**1667
Blood transfusion
(animal-to-human)**
Jean -Baptiste Denys, *Professor of
Philosophy and Mathematics at
Montpellier University, France, and
personal physician to King Louis
XIV. A lamb's blood was given to a
fifteen-year-old boy, who recovered.*
Wind gauge
Christian Förner *of Germany, to
measure the air pressure inside the
bellows of a church organ.*

**1668
Cast-iron pipeline**
*for supplying water to Versailles. It
stretched for 5 miles between the
reservoirs of Picardy and
Montbauron.*
Reflecting telescope
Isaac Newton, *mathematician, while
Fellow of Trinity College, Cambridge,
England.*

**1669
Phosphorus**
Hennig Brand, *alchemist, Germany.*

**1670
Megaphone**
Sir Samuel Morland, *'master
mechanic' to King Charles II of
England. Manufactured by Simon
Beal, Suffolk Street, London.*
Roberval Balance
Gilles Personne de Roberval,
mathematician and engineer, France.
Wheel-cutting machine
Robert Hooke, *Curator of
Experiments to the Royal Society,
London. To replace the laborious
process of cutting and filing the wheels
and pinions by hand.*

**1671
Spun silk**
Edmund Blood, *England – he
developed a machine for carding and
spinning silk.*

**1672
Hose-pipe (fire-fighting)**
Nicholas *and* Jan van der Heijden *of
Amsterdam, Holland. They devised a
suction hose to draw water out of the
canals and into a funnel, a gravity
system to feed the pump and a pressure-
hose to direct the water at the fire.*
Weather barometer
Otto von Guericke, *engineer-
showman, who mounted his 34 ft-high
brass tube containing water with a
closed glass section on the side of his
Magdeburg house so that passers-by
could see the little figure of a man
floating high on the water in good
weather, and sinking when the weather
was stormy.*

**1673
Artificial horse
(equestrian teaching machine)**
simultaneously by John Wells *and*
Raphael Folyart, *both of London.*

**1674
Musical time-signature**
Giovanni Maria Bononcini, *composer
and violinist in the Court Orchestra of
Modena, Italy.*
Tourniquet
Morel, *a French surgeon, to aid
amputation.*

**1675
Crystal-clear flint glass**
George Ravenscroft, *England,
commissioned by the Glass Seller's
Company to make better glass than the
Venetians. In 1676 Ravenscroft's
raven head seal appeared on the
Company's glassware.*
**The foot-rule
(measuring instrument)**
unknown inventor. A wooden 2-ft rod

calibrated in inches and subdivided
into eighths of an inch.
Spiral balance spring
Christiaan Huygens. *Incorporated in
a timepiece by Isaac Thuret of Paris.*
Water filtration system
William Woolcott, *England.*

1676
**Repeating clock (rack-and-snail
striking mechanism)**
Reverend Edmund Barlow *residing
at Parkhall, Lancashire, England.*
Universal joint
Robert Hooke, *Curator of
Experiments to the Royal Society,
London. To manipulate his helioscope
or safe, sun-observing machine.*

1677
**'Glassis' (very hard floor
plaster)**
Kenrich Edisbury, *of Hammersmith,
Middlesex, England.*

1678
Mechanical shuttle change loom
M. de Gennes, *France.*

1679
Amputation (flap method)
Lowdham *of Exeter, England.*

1680
'Anchor' escapement
William Clement *of London, who
made the first practical long pendulum
clock with the new escapement. But
Robert Hooke claimed that with the
assistance of Thomas Tompion, he had
invented the anchor pallet in 1666, the
year of the Great Fire of London.*
Lithotomy (lateral operation)
Franco.
Pressure cooker
Denis Papin, *French assistant to the
Hon. Robert Boyle in London. Papin's
'steam digester' was a container with a
tightly fitting lid, which increased the
pressure inside and considerably raised
the boiling point of water. The safety*

*Denis Papin's 'steam digester', the original
pressure-cooker*

*valve on the lid was also invented by
Papin, who claimed that his 'bone
digester' could make the oldest and
hardest cow-beef 'as tender and
savoury as young and choice meat'.*

1681
**Combined grenade discharger
and musket**
John Tinker, *England.*

1682
Brine-pumping machine
William Marbury *of Cheshire,
England.*
Halley's Comet
Edmund Halley, *astronomer and
mathematician at Oxford, England.
He subsequently decided that it must be
identical with comets previously seen in
1607 and 1531. He accurately
predicted its return for 1758. It has
been seen since in 1835, 1910 and 1986.*

1683
'Animalcules' (little animals)
Antonie van Leeuwenhoek, *a
secretive former cloth warehouse clerk
of Delft, Holland. He observed these,
using meticulously ground, high-
powered lenses (× 2–300*

A   B

C

Van Leeuwenhoek's 'animalcules'

multiplications) on saliva, teeth-scrapings, cow dung, etc. Unwittingly he had discovered bacteria.

Croissant (crescent-shaped bun)
Kulyezisk, *Polish baker in Vienna.*

Ironware-manufacturing machinery (mill-powered)
William Palin *and* William Loggins *of London.*

Sawing engine (for use by the disabled)
John Booth, *merchant, England.*

1684
Geometric-block communications
Robert Hooke. *Boards of different shapes were hung up in a large, square frame divided into four compartments. The characters could be varied in 10,000 ways and observed through telescopes. The same character might be seen in Paris the minute after it was shown in London. But the President of the Royal Society made the criticism that its use would be hindered by fogs and mists.*

Gravitational theory
Isaac Newton, *Lucasian Professor of Mathematics at Cambridge University, England. His book 'Principia' was published in 1687, but did not include the initial inspiration: an apple falling.*

Thimble
Nicholas van Benschoten, *Amsterdam.*

1685
Iron pens.

1687
Statistics
Sir William Petty, *economist, England.*

1688
Cast plate glass
Bernard Perrot, *glazier of Orléans, France. Following casting and rolling, the plates had to be laboriously ground and polished by hand.*

Champagne
Dom Pierre Pérignon, *monk and cellarer at the Benedictine Abbey of Hautvilliers, Champagne, France.*

Continuous-pattern wallpaper
Jean-Michel Papillon, *interior designer, France.*

Experimental rockets (50–100 lb)
Friedrich von Geissler, *Berlin.*

Mural arc
John Flamsteed, *English clergyman and first Astronomer Royal, with his assistant Abraham Sharp for use at the Greenwich Observatory. Since 1676, using sextants, quadrants, clocks and telescopes, Flamsteed had been measuring the motions of the heavens and locating the positions of over 2000 stars by means of tens of thousands of observations. It was not until 1725, six years after his death, that Flamsteed's epic catalogue was published.*

1690
Clarinet
Johann Christoph Denner, *instrument-maker of Nuremberg. Originally called the 'chalumeau'.*

Hydrapsis (water-shield)
Johannes Christoph Wagenseil *of Nuremberg 'as a machine by the help of which a person may walk upon the water without fear of sinking'. King Christian V of Denmark tried the hydrapsis and went more than a mile with it on the open sea.*

Light (finite velocity)
Ole Rømer, *Royal Mathematician and Professor of Astronomy at Copenhagen, Denmark.*

Denis Papin

Steam engine (experimental)
Denis Papin, *French Huguenot physicist working in exile at Marburg University. Dogged by bad luck, Papin ultimately died in poverty and melancholy in the slums of London.*

1691
Rollers to replace carriage wheels
Kendrick Edisbury, *of Hammersmith, Middlesex, England.*

1692
'Fives' (ball game)
Thomas Samborne, *England.*
Languedoc Canal
Pierre Paul Riquet, *France. 150 miles long, with over 100 locks and a 515 ft long tunnel, it took 26 years to complete.*

1694
Pollen (in plants)
Rudolph Jakob Cammerarius, *German physician and botanist.*

1695
Bank notes
Issued by the Bank of England.
Epsom salts
Nehemiah Grew, *vegetable physiologist, England.*

1697
Pantaleon
Pantaleon Hebenstreit, *tutor at Maseburg, Germany, as a drum with tuned strings.*

1698
Inflated girdle
Francis Cryns, *who demonstrated his life-preserver at Bristol and Portsmouth and in front of King William III at Windsor. Priced at a guinea and a half.*
Steam-powered mining pump
Captain Thomas Savery, *English military officer. It was used with partial success in pumping water out of Cornish mines, and also into buildings.*

1699
Fire pump (portable)
Dumaurier Duperrier *of Paris. By 1722 there were 30 such engines in Paris.*

1701
Machine seed drill
Jethro Tull, *Oxford graduate and barrister, while running his father's Howberry Farm, near Wallingford, Oxfordshire, England. It sowed seeds in precise rows. The farm labourers went on strike in protest, fearing the efficiency of the machine.*

1702
Borax/Boron
Guillaume Homberg, *alchemist in the laboratory of Philip II, Duke of Orleans, Paris, France.*
Tidal pump
George Sorocold *for the London Bridge Water Company, River Thames, England.*

1703
Binary arithmetic
Gottfried Wilhelm Leibnitz *of Leipzig.*

Savery's steam-powered mining pump,
1698

1704
Jewelled watch
Niccolo Facio, *Italy*.
Prussian blue pigment
Diesbach, *Berlin colour*
manufacturer, and his assistant,
Dippel. *They managed to keep their*
manufacturing method a secret until
1724. At first it was used as an artist's
pigment and later on in the century as a
dye.

1709
Anemometer
Wolfius, *to measure the strength and*
velocity of the wind.
Coke smelting (iron)
Abraham Derby *at Coalbrookdale,*
Shropshire, England. He had recently
come from Bristol to take over an
existing charcoal blast furnace set up in
1638.
Hammerklavier
Bartolommeo Cristofori, *harpsichord*
builder in Florence. Gottfried

Silbermann, *a German organ builder, further developed Cristofori's idea.*

1711
Tuning fork
John Shore, *Sergeant Trumpeter to the Court of Queen Anne of England and lutenist at the Chapel Royal.*

1712
Steam engine piston
Thomas Newcomen, *Dartmouth ironmonger and Baptist preacher, with his friend* John Calley, *a plumber and glazier. Their first practical pumping station was at Tipton, Staffordshire to pump water from the Earl of Dudley's Coneygre Colliery.*
Swell pedal (for church organs)
Abraham Jordan *of London. First box built at St Magnus the Martyr, London Bridge.*

1714
Fahrenheit scale
Gabriel Daniel Fahrenheit, *modest German physicist residing in Holland. He made important improvements in the construction of thermometers and introduced the thermometric scale known by his name by choosing zero as the lowest temperature obtainable with a mixture of common salt and ice.*
Writing machine
Henry Mill, *England. Machine and patent lost.*

1715
Dead beat escapement (clock device)
George Graham, *engineer, London.*
Diving suit (practical)
Andrew Becker. *Demonstrated in the River Thames, England.*

1716
True porcelain
Johann Friedrich Böttger, *German alchemist in the Royal Laboratory of Friedrich August II, Elector of Saxony. In 1711 Böttger came across* kaolin and felspar being sold as part of the wig industry. From these he made true 'hard paste' porcelain. To maintain secrecy, Friedrich the Strong compelled the alchemist to work in private, under an armed guard. By 1716 Meissen porcelain had been placed on the market.

1717
Diving bell
Sir Edmond Halley, *astronomer and mathematician.*
Perpetual motion drum
'Offyreus', *alias* Johann Ernst Elias Bessler. *'Tested' over two months in a carefully locked room in the castle of Charles, Landgrave of Hesse-Kassel.*

1718
Machine-gun (flintlock)
James Puckle, *London lawyer, for use on board ships. At a public demonstration in 1722, a Puckle gun discharged 63 bullets out of its eleven pre-loaded chambers in 7 minutes. Puckle's gun, manufactured at White Cron Alley factory, was found to be too heavy to be practical. It was designed to shoot round bullets against Christians and square bullets against Turks.*

1719
Full-colour printing process
Jakob Christof Le Blon *of Frankfurt, Germany.*

1720
Optical harpsichord
Louis Bertrand Castel. *Coloured tapes were fitted to each finger key so that, as light passed through these during a recital, a coloured light show was also enjoyed. Called the 'clavecin oculaire'.*
Pedal harp
unknown instrument-maker *in Germany.*
Types cast
William Caslon, *London gun engraver*

and toolmaker. He cut his first 'founts' in 1720 and the first specimen of printed work appeared in 1734.

1725
Stereotyping
William Ged, *Scottish printer and former goldsmith, Edinburgh. Developed a prototype built by Van der Mey of Leyden 200 years before.*

1726
Compensated mercury pendulum
George Graham, *watchmaker, London.*

1728
'Temple spectacles' (with rigid side pieces)
Edward Scarlett.

1729
Aberration of light
Reverend James Bradley, *Savilian Professor of Astronomy, Oxford University, England.*
Tensile-testing machine
Pieter van Musschenbroek, *Physics Professor at Utrecht, Holland.*

1730
Rotherham plough
Joseph Foljambe *of Holland. It was adapted from a Dutch plough with a wooden mouldboard.*
Thermometer (spirit)
René Antoine Ferchault de Réamur, *scientist and Director of the Paris Academy of Sciences – after whom the 80-degree scale is named.*

1731
Octant
John Hadley, *mathematical instrument maker of East Barnet, London. Known as 'Hadley's Quadrant'.*
Pyrometer
Pieter van Musschenbroek, *Physics Professor at Utrecht, Holland.*

1732
Copper-zinc alloy
Christopher Pinchbeck, *London clockmaker.*
Pitot Tube
Henri Pitot, *French physicist.*
Threshing machine
Michael Menzies *at Edinburgh. It was driven by a water wheel.*

1733
Arsenic
Georg Brandt, *Professor of chemistry, Uppsala University, Sweden.*
Child's perambulator
William Kent *for William Cavendish, 3rd Duke of Devonshire, England.*
Composite glass lens (crown-flint)
Chester Moor Hall, *Essex, England.*
Flying shuttle
John Kay, *woollen-mill owner's son and clockmaker at Colchester, England. Kay's invention was plagiarised by the Yorkshire clothiers who formed the 'Shuttle Club' to cover each other's costs should they be prosecuted. In a succession of (successful) lawsuits, Kay bankrupted himself and was pursued by angry weavers from Leeds and so to France, where he died in debt.*

1734
Fire-extinguisher bombs
M Fuchs *of Germany. Water-filled glass balls were thrown on the fire.*
Metal-plate rolling machinery (water-powered)
Kristofer Polhem, *Swedish scholar, engineer and industrialist.*

1735
Marine chronometer
John Harrison, *a self-educated Yorkshire carpenter, as an entry for a £20,000 prize offered by the British Admiralty's Board of Longitude for a clock which would have an error of less than 2 minutes on a voyage to the West Indies and back.*

Platinum
Don Antonio de Ulloa de la Torre
Giral, *mariner, Spain.*

1737
Contour lines
Philippe Buache, *French
cartographer, on a chart of the English
Channel.*
Didot Points
François-Ambrose Didot, *French
printer working in Paris, to
standardise type sizes.*
Linnaean system (botany)
'Linnaeus', alias Carl Linné, *Swedish
botanist of Uppsala, resident in
Holland.*

1738
**Automaton duck (including
digestive system)**
Jacques Vaucanson *of Paris.*

1739
Jewish Year 5500.
Pile-driving machine
Charles Valoue, *clockmaker, London.*

1740
Curare drug
Charles Marie de Lacondamine,
*French geographer, during a visit to
Peru and the Amazon. It was
originally used by the Peruvian
Indians on their arrow tips.*
Dividing engine
Henry Hindley *of York, England, for
cutting the teeth of clock wheels.*
Microscope (pocket-sized)
Benjamin Martin, *instrument maker,
Chichester, Sussex, England.*

1741
Fizzy mineral water
Dr William Brownrigg, *chemist, of
Whitehaven, Cumbria, England.*
Ventilators
Reverend Stephen Hales, *perpetual
curate of Teddington, Middlesex,
England. He also invented machines*
*for distilling sea-water, preserving
meat, etc.*

1742
Ballistic pendulum
Benjamin Robins, *English
mathematician. This was the first
accurate means of testing the force of
gunpowder.*
Celsius scale
Anders Celsius, *Swedish astronomer
at Uppsala – to express temperatures
on a Delisle thermometer as degrees (°C)
on the International Practical
Temperature Scale. Celsius's scale is
believed to have been invented by his
compatriot, Carl Linné.*
Crucible steel technique
Benjamin Huntsman, *a clockmaker
from Doncaster, England.*
Indoor swimming pool
'The Bagnio', Lemon Street,
Goodman's Fields, London.

1743
**Carding machine (wire-toothed
revolving cylinders)**
David Bourne, *Leominster,
Herefordshire, England.*
**Cart weighbridge (compound
lever)**
John Wyatt, *carpenter, Staffordshire,
England.*
'Chaise volante' (elevator)
*installed in the private apartments of
King Louis XV of France in the Petite
Cour du Roi at Versailles, to give the
king access to his mistress, Madame de
Châteauroux, on the floor above. It
was worked by weights and pulleys.*
Sheffield plate
Thomas Boulsover *of Sheffield,
England. It was originally used for the
manufacture of buttons.*
Whist (card game)
Edmund Hoyle, *England.*

1745
Windmill (automatic fantail)
Edmund Lee, *England. To keep the
sails square to the wind.*

1746
Leyden jar (prototype electrical condenser)
Simultaneously by Pieter van Musschenbroek, *Dutch scientist, of Leyden and* Dean E G von Kleist *of Cammin, Pomerania.*

1747
Boxing gloves
Jack Broughton, *London pugilist.*
Diet deficiency
James Lind, *British naval surgeon on board* HMS *Salisbury, who discovered that by making sailors suffering from scurvy eat a diet of oranges and lemons, there was at once a return to good health. Lind went out of his way to bring this cure to the attention of the British Admiralty.*
Multiple-mirror solar furnace
Count Louis Leclerc Buffon. *Using between 100 and 360 small flat mirrors, individually pointed to send sunlight to a common focus, Buffon was able to ignite woodpiles over 60 metres away. The same year, using a burning glass 112 cm in diameter, Jacques Cassim of the Paris Observatory was able to melt silver to such a fluid state that the drippings formed hairlike strands when chilled in water.*

1748
Sea quadrant
B Cole *in the Orrery, Fleet Street, London, England.*

1750
Papier-mâché
John Baskerville, *printer of Birmingham, England, who also designed 55 typefaces.*
Prony brake (dynamometer)
Gaspard de Prony, *France.*
Wove paper
John Baskerville, *using woven wire sieves to produce the 'laid' effect.*

1751
Nickel
Axel Fredrik Cronstedt, *metallurgist, Swedish Academy of Sciences.*
Planing machine (for iron)
N Focq, *France.*

1752
Geometrical pen
G Suardi, *Italy.*
Lightning conductor
Benjamin Franklin, *for his own house in Market Street, Philadelphia, Pennsylvania. His steel-pointed, 7 ft iron rods were also mounted on the State House and the Academy Building during the same year. Their erection followed experiments with moistened silk kites and iron keys to assess electric discharge from thunderstorms.*

Benjamin Franklin

Parabolic reflector
William Hutchinson, *marine consultant, England.*

1753
Writing machine
Professor Friedrich von Knauss, *Director of the Physical and Mathematical Laboratory, Vienna.*

1754

1754
Lightning conductor
'Procopius Divis', *Moravian monk,*
independently of Benjamin Franklin's
experiments.

1755
Bridge (iron-girdered)
M Garvin, *French engineer. Of a 25*
metre span across the River Rhone,
only one of the three arches was of
forged iron, the remaining two being of
wood.
Kaolin
William Cookworthy, *druggist and*
Quaker at St Austell, Plymouth,
England.

1756
Carbon dioxide
Joseph Black, *chemist, Glasgow*
University, Scotland.

1757
Sextant
Captain John Campbell, *as a*
navigational improvement on John
Hadley's quadrant of 1731.

1758
Achromatic lens combination
John Dollond, *Huguenot silk weaver*
turned London optician. The soda-lime
type of lens was used for crown-glass
and the potash-lead type for flint glass.
Achromatic refractor telescope
John Dollond.
Aneroid barometer (prototype)
Zaiker, *for use at sea.*
Revolving stage
Namiki Shozo *for the Kado-za doll*
theatre in Osaka, Japan.
Ribbed stocking-frame machine
Jedediah Strutt, *former Blackwell*
farmer, on the suggestion of his
brother-in-law.

1759
Lighthouse (Portland stone and
hydraulic cement construction)
John Smeaton, *civil engineer and*

former instrument maker, England.
The third 'Eddystone' lighthouse.

1760
Cacodyl
Louis Claude Cadet de Gassicourt.
Described as 'one of the worst smells in
chemistry'.
Cast-iron cog wheels (for mills)
Carron Iron Works, *Glasgow,*
Scotland.
Parallelogram window blinds
Dr Gowin Knight, *scientist, London.*
Roller-skates
Joseph Merlin *of Huy, Belgium, a*
violinist with an ill-fated sense of
showmanship.
Screw-manufacturing machine
Job *and* William Wyatt, *England.*
'The Young Writer' (automaton
child doll)
Pierre Jaquet-Droz, *Swiss*
watchmaker. It could write a letter of
50 words.

1761
Clavecin électrique (electric
harpsichord
Jean Baptiste Delaborde *of Paris. It*
was unsuccessfully powered by Leyden
jars.
Marine chronometer (pocket-
watch size)
John Harrison, *Yorkshire carpenter.*
During a voyage to Jamaica, it lost 5
seconds during 81 days at sea. But it
was only through the personal
intervention of King George III that
the British Admiralty paid Harrison
the prize money they had originally
offered in 1713.
Percussion (medical diagnostic
technique)
Joseph Leopold Avenbrugger,
physician, Spanish Hospital, Vienna,
Austria.
Scissors and shears (mass-
produced in cast steel)
Robert Hinchliffe *of Sheffield,*
England.
Vaulting horse
Lund University, Sweden.

1762
Brewing thermometer
Michael Combrune, *France.*
Fire-extinguisher
Dr Godfrey *of London. Sal
ammoniac-filled containers were burst
by gunpowder.*
Latent heat
Joseph Black, *Professor of chemistry,
Glasgow University, Scotland.*
Rubber solvent
François Fresnau *at Marennes,
France.*
Sandwich
John Montagu, *4th Earl of Sandwich,
during a 24-hour gaming session in
London.*

1763
Cork lifejacket
Dr J Wilkinson. *It was accepted by
the British Admiralty for use in the
Fleet.*
Jigsaw puzzle
John Spilsbury *of Drury Lane,
London, 'Engraver and Map Dissector
in wood', to facilitate the teaching of
geography.*
Printed cheque
Hoare's Bank, *London.*
Queen's Ware pottery
Josiah Wedgwood, *English potter. He
bought from several inventors a process
which drastically reduced the time and
cost of producing decorated
creamware. He also invented
Jasperware.*

1764
Bolting machine
John *and* Thomas Morris, *hosiers of
Nottingham, England, were financed
by* John *and* William Betts.
Road-making system
Pierre Trésaguet *of the Corps des
Ponts et Chaussées. He developed a
base of flat stones set vertically in a
rough voussoir-arch shape – later
adopted by Thomas Telford in Great
Britain.*

Spinning jenny
James Hargreaves, *illiterate weaver
and carpenter of Standhill, near
Blackburn, England. The number of
spindles on his first machine was eight,
reproducing the actions of eight hand-
spinners. Later it rose to twenty or 30
and eventually 120. Whether the word
'jenny' refers to his little daughter or
his wife, or is a nickname for the
engine, is unresolved.*

1765
Hard porcelain (French system)
Louis-Félicité de Brancas, Comte de
Lauraguais, *using kaolin or china-clay
from Alençon in France. Three years
later,* Pierre Jean Macquer *identified
kaolin at St Yrieix-la-Perche.*
Steam engine condenser
James Watt, *Glasgow instrument
maker, originally commissioned to
repair a small Newcomen engine. Watt
made his engine more fuel-efficient by
producing a separate condensation
chamber. Together with a Birmingham
manufacturer called Matthew
Boulton, 'Jimmy' Watt changed the
face of the industrial world, their Soho
factory turning out hundreds of steam
engines.*

1766
Fire escape
David Marie, *watchmaker,
Westminster, London.*
Glass harmonica
Benjamin Franklin, *American
ambassador and scientist, living in
London. A treadle-operated 'vérillon'
with self-tuned glasses. Following the
addition of a keyboard in 1785, both
Mozart and Beethoven wrote for it.*
Hydrogen (gas)
Henry Cavendish, *35-year-old
natural philosopher, living at
Clapham Common, South London.
His discovery of what he called
'inflammable air' led to early balloon
experiments – as hydrogen is 10.8 times
lighter than common air.*

1768

Lightning conductor placed on top of St Mark's, Venice.

Mariner's direct-reading compass Dr Gowin Knight. *Better steel for the needle for increased magnetic strength, better support for the compass bowl and a gimbal system were the features of the Knight compass as adopted by both the British Royal Navy and Germany.*

Mesmerism Friedrich Anton Mesmer *and* Professor Hehl, *a Jesuit priest and astronomer in Vienna, from their interest in animal magnetism.*

Watercolour pigment cakes William *and* Thomas Reeves *of London. They were rectangular in shape, and by 1781 there was a choice of 40 colours.*

1768

Areometer Antoine Baumé, *master apothecary, Paris. For measuring specific gravity of fluids.*

Dividing engine Jesse Ramsden, *instrument maker, Haymarket, London.*

1769

Automaton chess player Baron Wolfgang von Kempelen. *His automaton was dressed as a Turk and the Baron had people believe that it was all done with levers, gears, drums and cylinders. In fact it was done by a confederate 'out of sight' using a system of balls and magnets. Emperor Joseph II of Austria, Empress Catherine of Russia and even Napoleon were convinced it was an automaton.*

Lifejacket L'Abbé de La Chapelle, *French priest.*

Model boat (steam-powered) William Henry, *gunsmith of Lancaster, Pennsylvannia. It was tested on the Conestoga Creek.*

Tartaric acid Carl Wilhelm Scheele, *chemist in*

Stockholm, Sweden.

Venetian blinds Edward Bevan, *England.*

Water frame Richard Arkwright, *former wig-maker and hair-dyer of Bolton, England, with* John Kay, *Warrington clockmaker. While their first spinning frame mill at Nottingham (1771) was horse-powered, a second plant at Cromford, Derbyshire was powered by the River Derwent – hence the term, water frame.*

Water wheel shaft (cast iron) John Smeaton *civil engineer, Austhorpe, England.*

1770

Electric battery (135 Leyden jars) John Cuthbertson, *English instrument maker in Amsterdam, Holland.*

Eraser (india-rubber) Mr Nairne, *mathematical instrument maker, the Royal Exchange, London.*

False teeth (mineral paste) A Duchâteau, *apothecary, France.*

Nail violin Johannes Wilde, *St Petersburg, Russia. Metal nails instead of strings.*

Phrenology Dr Franz Joseph Gall, *German-born medical student in Vienna. He began with his schoolfellows, then studied the heads of criminals and others, delivering his first lecture in Vienna in 1796. In 1802 the Austrian government prohibited his teaching. He was joined by* Dr Johann Caspar Spurzheim *and their researches led to an increased study of the brain.*

Ramsden's lathe Jesse Ramsden, *Halifax-born instrument maker, Haymarket, London.*

Steamcar Nicholas Joseph Cugnot, *French artillery officer, whose 'voiture en petit' built for the National Arsenal could easily carry four people at 2½*

mph though it had to stop every 15 minutes for the boiler to be re-filled by hand.
Sulphur dioxide
Joseph Priestley, *chemist and Presbyterian minister, Mill Hill Chapel, Leeds, England.*

1771
Fluorine
Carl Wilhelm Scheele, *chemist, Uppsala, Sweden.*
Picric acid
Woulfe.

1772
Automaton musician
Henri-Louis Jaquet-Droz *and his associate* Leschot. *The lady played five tunes on the organ, with a curtsy between each tune.*
Coloured inks patent.
Weighing machine with dial
John Clais, *London.*
'Empyreal air'
Carl Wilhelm Scheele, *chemist, Uppsala, Sweden.*
Nitrogen ('noxious air')
Daniel Rutherford, *Scottish physician and botanist.*
Printing press (cast-iron)
Wilhelm Haas, *typefounder at Basle.*
Varnish formulary
Jean Felix Watin, *France.*
Windmill-spring sail
Andrew Meikle, *millwright, Houston Mill, Dunbar, Scotland. His hinged shutters replaced sailcloths.*

1773
'The Designer' (automaton draughtsman)
Henri-Louis Jaquet-Droz, *Swiss automaton-maker, and his associate* Leschot.

1774
Ammonia
Joseph Priestley, *chemist and Presbyterian minister at Mill Hill Chapel, near Leeds, England.*

Barium
Carl Wilhelm Scheele, *chemist, Uppsala, Sweden.*
Chlorine gas
Scheele.
'Dephlogisticated air'
Joseph Priestley.
Manganese
Scheele.
Tea fountain
John Wadham, *England.*

1775
Borer and polisher (steam-engine cylinders)
John Wilkinson, *Bersham Furnace, Wrexham, England. Based on Smeaton's prototype of 1765.*
Chain-driven machine
Crane, *England. Used to manufacture ladderproof stockings.*
Digitalis (as drug)
William Withering, *botanist and chief physician of Birmingham General Hospital, England.*
Electrophorus
Count Alessandro Volta, *Italian physicist and Professor of Natural Philosophy at Pavia. For obtaining friction electricity.*
Pressure anemometer
Dr James Lind, *Scottish physician, Windsor, England. For measuring extreme velocity of the wind at 93 mph.*
Valve water closet (S-shape pipe)
Alexander Cummings, *watch and clockmaker of Bond Street, London. Modified three years later, this WC sold 6000 units during the next two decades.*

1776
Automatic lace-making machine
Leture, *French engineer.*
Breech-loading rifle
Colonel Patrick Ferguson, *Scotland.*
Submersible
David Bushnell, *Saybrook, Connecticut, USA. The 'Turtle' was*

submerged by taking water into its tanks and raised by pumping it out again. Hand-cranked by Sergeant Ezra Lee, 'Turtle' attempted the first submarine attack on Lord Howe's flagship 'Eagle'. Unable to fix an explosive charge on 'Eagle's' bottom, 'Turtle' had to make a quick getaway.

Threshing machine
Andrew Meikle, *millwright, Houston Mill, Dunbar, Scotland.*

Torsion balance
Charles Augustin de Coulomb, *French scientist working in Martinique.*

1777
Circular saw
Samuel Miller, *England. It did not come into regular use until the application of steam power, after which it sold in numbers around the sawmills of Great Britain and the Continent.*

Harecastle Tunnel
James Brindley, *to carry the Trent and Mersey Canal under Harecastle Hill near Stoke-on-Trent, England. Just wide enough for a narrowboat, it was 2926 yd long and took eleven years to construct.*

Iron boat
12 ft pleasure craft on the River Foss, Yorkshire, England.

'Oxygine' (Greek, acid maker)
Antoine Laurent de Lavoisier,

Lavoisier's apparatus for analysing 'Atmospheric Air'

French chemist and director of the government powder mills.

1778
Copying machine
James Watt *at the Soho Works, Birmingham, England. 150 machines were sold before the end of the year, and it was patented in 1780.*

Molybdenum
Carl Wilhelm Scheele, *chemist and apothecary, Köping, Sweden.*

Mortise lock (many-tumbler)
Robert Barron, *locksmith of Shoreditch, London.*

1779
Carronade (naval gun)
General Robert Melville, *cast by the Carron Company of Falkirk, Scotland.*

Cast-iron bridge
'Ironbridge' over the River Severn at Coalbrookdale in Shropshire, England.

Glycerine
Carl Wilhelm Scheele, *chemist and apothecary, Köping, Sweden.*

Rotative steam engine
Matthew Wasborough *of Bristol, England. It was used to power machinery at James Pickard's button manufactory in Birmingham.*

Spinning mule
Samuel Crompton, *26-year-old farmer's son from Firwood, near Bolton, Lancashire, England, who was only granted government compensation for his invention – some £5000 – when aged 59.*

1780
Argand lamp (circular wick)
Aimé Argand, *Swiss physician, chemist and balloonist. A patent for this lamp was granted in 1784 but its validity was disputed and in 1787 Argand died in poverty at Versoix, near Geneva.*

Artificial insemination
Lazzaro Spallanzani, *Italian priest*

Carronade, 1779

and biologist working in Pavia.
Lighthouse (glass reflectors).
Pyrometer
Josiah Wedgwood, *pottery*
manufacturer of Hanley, England.
Two-horse swing plough
James Small, *Scotland.*
War rockets
Hyder Ali, *ruler of Mysore, India,*
against the British at Guntar.

1781
Compound steam engine
Jonathan Carter Hornblower,
Cornish mining engineer, Radstock
Colliery, Bath, England.
Eidophusikon
Philip James de Loutherbourg,
French-born stage- and scenery-
designer for David Garrick, the actor.
Natural phenomena represented by
moving pictures. Exhibition at Lisle
Street, Leicester Square, London.
Hair hygrometer
Horace Bénédict de Saussure,
physicist and geologist, Professor of
Philosophy, Geneva, Switzerland.
Sun-and-planet gearing system
(steam engines)
William Murdoch, *engineer-assistant*
to James Watt, Soho Works,
Birmingham, England.

Uranus (planet)
Frederick William Herschel,
German-British astronomer, former
oboist in the Hanoverian Guards band
and former organist and music teacher
in fashionable Bath, England. It was
first called 'Georgium Sidus' after
King George III, next 'Herschel' and
finally Uranus.

1782
Tellurium
Franz Joseph Müller, *chemist,*
Director of the State Mines of
Transylvania, although given its name
by Martin Heinrich Klaproth.

1783
Balloon
Joseph-Michel *and* Jacques-Etienne
Montgolfier, *paper-makers of*
Annonay, near Lyons, France, who
gave the first public demonstration of
their 'hot-air' balloon (of 110 ft
circumference) in their town square. It
was made of cloth with paper and
fastened together with buttons.
Following the aerial flight of a cock, a
sheep and a duck suspended in a basket,
the first men-carrying balloon, the
'Montgolfière', ascended from the Bois
de Boulogne and drifted some 5 miles
across Paris.

Balloon (hydrogen-filled and valve-controlled)
Jacques Alexandre Charles, *Parisian physicist, and the* Robert brothers. *Its first flight was 27 miles.*

Centrifugal governor for windmills
Thomas Mead, *carpenter of Sandwich, Kent, England.*

Cotton printing (cylindrical roller process)
Thomas Bell, *Scotland.*

Paddleboat (steam-engined)
Marquis Claude de Jouffroy d'Abbans *and* Comte Charles de Follenay. *Equipped with an engine built by Frères Jean of Lyons, France, 'Pyroscaphe steamed several hundred yards against the current of the River Saône at Lyons.*

Pianoforte (sustaining and damper pedals)
John Broadwood, *London cabinet-maker.*

Wolfram (tungsten)
Don Fausto d'Elhuyar, *Spanish chemist, and his brother,* Juan José.

1784
Bifocal lenses
Benjamin Franklin, *while* US *minister in Paris.*

Gas-lit room
Professor Jean Pierre Minckeler *in the lecture-room of the University of Louvain, France.*

Mail coach
John Palmer, *theatre manager of Bath and Bristol, England.*

Model helicopter (twin rotors)
Launoy, *French academician and naturalist, and his mechanic Bienvenue. Demonstrated to the Academy of Sciences.*

Puddling furnace (iron)
Henry Cort, *ironmaster of Funtley, Hampshire, England, and also a navy agent in London. He also invented a prototype rolling mill.*

Rope-making machine
Richard March, *hosier, London.*

Saccharometer in brewing
John Richardson *of Hull, England.*

Safety combination lock
Joseph Bramah, *lame Yorkshire mechanic. Bramah was so confident about his burglar-proof barrel-shaped lock, with its 494 million combinations, that he offered a prize of 200 guineas to the first person who could pick it. The prize remained unclaimed for 67 years until a* US *locksmith, Alfred C Hobbs, picked the lock after one month's work.*

Shrapnel shell
Lieutenant Henry Shrapnel, *Royal Artillery, England. His shell, filled with tiny lead balls, was designed to burst in mid-air and disperse metal fragments into the bodies of the enemy. It was adopted by the British Artillery in 1803.*

Steam winding colliery engine
a Boulton & Watt engine was adapted at the Walker Colliery on Tyneside, England.

Windmill-powered carriage
Bishop W Wilkins MA. *Hand-cranked to turn a four-bladed 'windmill' puller propeller.*

1785
Beer-pump handle
Joseph Bramah, *London.*

Grist mill factory (fully automatic)
Oliver Evans *of Philadelphia,* USA.

Human artificial insemination
M Thouret, *doyen of the Medical Faculty at Paris University. An intravaginal injection of sperm on his sterile wife with a tin syringe resulted in the birth of a healthy baby.*

Methane and ethylene
Comte Claude Louis Berthollet, *France.*

1786
Breech-loading musket
Henry Nock *of Ludgate Hill, London.*

Barbeu du Bourg's 'lightning conductor umbrella'

Lightning-conductor umbrella
Barbeu du Bourg, *having recently translated the writings of Benjamin Franklin into his native French.*
Match (phosphoric or philosophical bottle technique).
Raised letters for the blind
Valentin Haüy, *founder of the National Institute for Blind Youth, Paris, using embossed Roman letters.*
Steamboat (endless paddlechain)
John Fitch, *gunsmith of Windsor, Connecticut,* USA. *The Philadelphia-built hull was propelled by twelve vertical paddles, steam-operated by Fitch's own engine. Bankrupted in an attempt to interest the French*

government *in his invention, Fitch worked his passage back to the* USA *and later poisoned himself.*

1787
Eau de Javel
Comte Claude Louis Berthollet, *French chemist, for bleaching clothes.*
Flax-spinning machinery
Kendrew, *an optician, and* Porthouse, *a clockmaker, both of Darlington, England.*
Iron ship
John Wilkinson's *70 ft barge, 'Trial'.*
Power loom
Dr Edmund Cartwright, *Rector of Goodby, near Marwood, Leicestershire, England. The prototype was built by both the local carpenter and smithy, then powered by two strong men. Attempts to employ steam-engine versions at Doncaster and Manchester (400 power looms) met with fierce opposition and arson.*
Roller-bearings
John Garnett, *merchant of Redland, Gloucestershire, England.*
Steamboat (waterjet-propelled)
James Rumsey, US *engineer. Travelled upstream on the River Potomac.*
Theodolite
Jesse Ramsden, *Fellow of the Royal Society, and instrument-maker for the Trigonometrical Survey of England and Wales.*

1788
False teeth (wax mould and plaster model)
Dubois de Chemant, *Paris dentist.*
Panorama
Robert Barker, *Irish portrait painter. Paintings around the wall of a circular building. His first work was a view of Edinburgh.*
Phaeton (carriage).
Steamboat (multi-hulled)
Patrick Miller, *inventor, with* William Symington, *engine-builder, and* James Taylor. *It successfully*

steamed on Loch Dalswinton on
Miller's Dumfries estate, Scotland.

1789

Anémocorde
J J Schnell, *who took the then popular
Aeolian harp of ancient origin and
instead of allowing the natural wind or
breezes to produce music from its ten
strings, fitted a bellows and a
keyboard. The Polish Aeolopantaleon
of 1825 and Niccolo Isouard's 'piano
éolien' of 1837 were modified versions
of this.*
Bourbon (drink)
Eliza Craig, *Kentucky,
Massachusetts*, USA.
Mural circle (5 ft)
Jesse Ramsden, *English instrument-
maker, for Palermo Observatory,
Italy.*
Roller reefing sail (windmills)
Stephen Hooper, *England, who also
invented a horizontal windmill.*
Steam-engined power loom
John Austin, *Glasgow, Scotland. A
variation on Cartwright's loom of two
years before.*
Uranium (metal)
Martin Heinrich Klaproth, *Professor
of Chemistry at Berlin University,
Germany.*
Zirconium
Klaproth.

1790

Boot-sewing machine
Thomas Saint, *London cabinet-
maker.*
**Crane-neck phaeton (horse-
drawn carriage)**
'Highflyer', *built in London.*
Galvanic electricity
Professor Luigi Galvani *of Bologna,
Italy. His wife was skinning frogs with
a scalpel and was horrified to notice
that when the scalpel touched the frog's
salinated thighs resting on a zinc
platter, the thighs twitched. From this
incident, Galvani developed his theory
of 'animal electricity'.*

Ice yacht
Oliver Booth, USA.
**Machine for spinning and roving
cotton**
William Pollard, *Philadelphia. The
earliest* US *patent.*
**Tower clock (windmill-
powered)**
Benjamin Hanks, *New York*, USA.
**Two-wheeled wooden hobby
horse**
Le Comte de Sivrac, *Paris. Called
'Célérifère' (quick walker).*
Wool-combing machine
Revd Edmund Cartwright, *England.*
Wristwatch
Henri-Louis Jaquet-Droz *and*
Leschot, *Swiss watchmakers of
Geneva.*
Zinc white
Bernard Courteois, *French chemist.*

1791

Contour lines (land map)
Jean-Louis Dupont Triel, *French
surveyor.*
Gas-turbine (design)
John Barber *of Nuneaton, England.
Although fully patented, it was never
built.*
Guillotine
Joseph Ignace Guillotin, *French
physician, for efficiently inflicting
capital punishment by decapitation
during the French Revolution, thus
avoiding the cruel feudal use of the
wheel, the stake and the gibbet. At first
the machine was called the 'Louisette',
but was later nicknamed 'Madame La
Guillotine'.*
Ophicleide
Louis Alexandre Frichot, *French
instrument-maker working in London.
It was a keyed bassoon-cum-bugle and
was later replaced by the tuba.*
Titanium in ilmenite
William Gregor, *chemist, mineralogist
and Rector of Diptford, Devonshire,
England.*

1792
Ambulance
Baron Dominique Jean Larrey,
personal surgeon to Napoleon.
Larrey's Ambulance Corps, with
twelve fully manned spring-mounted
ambulances per division, was first used
in Napoleon's Italian campaign of
1796–7.
**'Cordelier' (rope-spinning
machine)**
Reverend Edmund Cartwright, *later*
improved by Captain Huddart.
Cottage (lit by coal gas)
William Murdoch, *Scottish-born*
area manager for Boulton & Watt
steam-engines in Cornwall, achieved
this at his Redruth home with help
from the local physician, Dr Boaze.
Graphite-clay pencil
Jacques-Nicolas Conté *of France,*
and independently by Josef
Hardtmuth *of Austria.*
Optical telegraph
Claude Chappe, *French mechanic,*
who perfected his T-type moving-arm
machine with the help of the noted
clock and watchmaker, Abraham
Louis Breguet. *Following its*
demonstration to the authorities, the
first Chappe Telegraph Line was set
up between Paris and Lille in 1794.
**Water-chemical fire
extinguisher**
von Ahen *and* Nils Moshein,
Sweden.

1793
Astigmatism
Thomas Young, *21-year-old medical*
student at St Bartholomew's Hospital,
London. For this discovery he was
made a Fellow of the Royal Society.
Daltonism (colour-blindness)
John Dalton, *teacher of mathematics*
and science in New College,
Manchester, England.
Public zoo
Le Jardin des Plantes, Paris.
Revolving stage (full-size)
Nakamura-za theatre, Edo, Japan.

Strontium
Thomas Charles Hope, *Professor of*
Chemistry, Glasgow, Scotland.
Veneer planer, shaver, etc
Samuel Bentham. *His brother,*
Jeremy Bentham, the famous writer on
political economy, was in charge of
several industrial prisons, whose
inmates were incapable of learning
manual trades. Samuel therefore
developed woodworking machines
which the prisoners could handle.
These included a veneer planer for
plywood panels, the first jig-saw,
circular saw tables, moulders, shapers
and lathes.

1794
Ball bearings
Philip Vaughan, *who also designed an*
axle using ball bearings.
Cam-operated weaving loom
William Bell *of Dumbarton, Scotland.*
Cotton gin
Eli Whitney, *Massachusetts*
University law graduate, with
Phineas Miller, *at Mulberry Grove*
cotton plantation, Georgia, USA. *By*
1800, US *cotton output had risen from*
140,000 lb to 35 million lb per year,
thanks to Whitney's gin.

Whitney's prototype cotton gin

**Machinery powered by
inflammable vapour**
Robert Street, *London varnish-
maker.*

Maximum and minimum thermometer
James Six, *mathematical instrument-maker, Maidstone, Kent, England.*
Safety curtain (theatre)
Drury Lane, London. Nicknamed 'The Iron'.
Yttrium
Professor Johann Gadolin, *Finnish chemist, in a mine at Ytterby, Sweden.*
The metal was first obtained by
Friedrich Wöller, *German chemist, in 1828.*

1795
Hydraulic press
Joseph Bramah, *England.*

1796
Carbon disulphide
Lampadius.
Lithography
Alois Sennefelder, *unsuccessful Bavarian actor. Although he originally used porous limestone, later on zinc plates became more popular.*
Smallpox vaccination
Dr Edward Jenner *at his native Berkeley, Gloucestershire, England, after twenty years' research. The first successful inoculation was given to eight-year-old James Phipps.*

1797
Alcohol engine
Dr Edmund Cartwright, *who combined a steam engine with a 'Scotch' fast-going still.*
Chromium
Louis Nicolas Vauquelin, *chemistry professor, School of Mines, Normandy, France.*
High beaver hat
John Hetherington, *haberdasher of London, who was arrested for creating a disturbance when Londoners rioted at the sight of such an innovation.*
Hydrotherapy (revival)
Dr James Currie, *Scottish physician practising in Liverpool, England.*

Orange marmalade (commercial production)
James Keiller & Son, *Dundee, Scotland. Based on Janet Keiller's recipe.*
Parachute
André Jacques Garnerin, *former balloon inspector in the French Army, who made his first descent from a balloon at a height of 2230 ft above Paris.*

1798
Charles's Law
Jacques Alexandre César Charles, *Professor of Physics, Conservatoire des Arts et Métiers, Paris.*
Gaslit factory
Boulton & Watt's Soho factory, Birmingham, England. Based on developments by their employee, William Murdoch.
Magnetic variometer
Henry Cavendish, *wealthy bachelor grandson of the 2nd Duke of Devonshire, also chemist and physicist. It was invented to record variations of the earth's magnetic field between two points.*
Mass production
Eli Whitney, *inventor, New Haven, Connecticut,* USA. *In order to make 10,000 muskets for the* US *Army in fifteen months, he subdivided the work into a great number of small jobs so that all the component parts of the musket could be made to such an accuracy as to be interchangeable. To eliminate the element of human imprecision, Whitney had the various milling, boring, grinding and polishing machines specially adapted.*
Optical glass manufacture
Peter Louis Guinard, *bell founder, Switzerland.*
Slide rest lathe
Henry Maudslay, *English toolmaker, Oxford Street, London.*
Transparent soap
Andrew Pears, *Cornish-born hairdresser working in Soho, London.*

1799

Clavicylindre
Ernst F F Chladni, *physicist of Wittenberg. Tuned glass rods were pressed against a rotating cylinder by means of a keyboard and thereby set in vibration.*

Enamel kitchen utensils
Samuel S Hickling, *Birmingham, England.*

Engine driven by condensed wind
George Medhurst, *England.*

Metric system
French Academy of Sciences, *after nine years' research. Committee members, Délambre and Méchain, measured an arc of the meridian between Dunkirk and Barcelona and from subsequent calculations they arrived at the metre in 1795.*

Paper-making machine
Nicholas-Louis Robert, *clerk at Pierre-Francois Didot's Essone Paper Works, France. Revolutionary France did not favour further development, so Didot took the invention to England.*

Pedometer
Ralph Gout.

Thermolampe (gas-fuelled)
Philippe Lebon *of Bruchay, near Joinville, France.*

1800

Cavalry sword (curved blade)
Colonel Jean Gaspard le Marchant, *France.*

Clock movement manufacturing machine (water-powered)
Eli Terry, *Plymouth, Connecticut, USA.*

Continuous current pile (prototype electric battery)
Alessandro Volta, *Professor of Natural Philosophy at Pavia, Italy.*

Infra-red region (light spectrum)
Sir William Herschel, *German-British astronomer.*

Iron-frame piano
John Isaac Hawkins, *English instrument-maker residing in Philadelphia, USA.*

Mechanically-produced pasta
Naples, Italy.

Paper (vegetable fibre)
Matthew Koops. *He developed a process for pulping straw, nettles, wood and even waste papers (recycled) to replace the system using old cotton linen rags. Koops's paper was, however, off-white hence not of the best quality.*

Screw-cutting lathe
Henry Maudslay, *English toolmaker, Oxford Street London.*

Screw-down tap (stop-cock)
Thomas Gryll, *Cornish-born ironmonger working in Birmingham, England.*

Seppings blocks (for suspending vessels in dock)
Robert Seppings, *master shipwright assistant, Plymouth Dockyard, Devon, England.*

Submarine (cigar-shaped)
Robert Fulton, USA. *The 'Fulton' had a pointed hull covered with copper plates and oxygen to support four men and two small candles for 3 hours. Propulsion was by hand-cranking and trials were made off New York.*

Tram-roads
Benjamin Outram, *civil engineer, constructed an iron tram-road from Croydon to Wandsworth, Surrey, England, which was completed in 1801.*

1801

Asteroids
Guiseppe Piazzi, *Italian astronomer, Director of the Palermo Observatory in Sicily. During the compilation of a star catalogue, he discovered Ceres, lying between Mars and Jupiter. This was the first of many new asteroids to be discovered.*

Engineering micrometer
Henry Maudslay, UK. *It was nicknamed 'The Lord Chancellor', for from it there was no appeal.*

Niobium
Charles Hatchett, *chemist,* UK. *It was originally called 'columbium'.*

Pulley-block-making machinery
Marc Isambard Brunel, *designer, and* Henry Maudslay, *toolmaker, to produce at least 100,000 pulley blocks for the Royal Navy during the Napoleonic crisis. By 1808 the plant was in operation in Portsmouth and the output of 130,000 blocks per year required ten unskilled men to do the work of 110 skilled craftsmen. A 30 hp steam engine powered the sequence of machinery. The government was saved £18,000 per year for a capital outlay of £54,000.*

Pumping station (steam-powered)
the Fairmount Water Works, *to supply Philadelphia,* USA, *with water.*

Semaphore
Dépillon, *former French artillery officer, with a three or four-arm machine. Dépillon's system furnished 301 signals, 105 more than the Chappe telegraph. By 1803 the Semaphore had been installed along the whole of the French coastline for ship-to-shore communication.*

Sugar-beet extraction factory
Franz Karl Achard, *Berlin chemist of Huguenot descent. Sited at Cunern, near Steinau in Silesia, the factory was sponsored by Frederick William III of Prussia.*

Wave theory of light
Thomas Young, *whilst working at Emmanuel College, Cambridge,* UK.

1802
Commercial steamboat
William Symington, *Scottish engineer for Lord Dundas of Kerse, to see whether steam could replace towpath-horse-towing. Tried on the Forth and Clyde canal, the 'Charlotte Dundas' was able to tow two 70-ton barges for six hours against a headwind.*

Perpetual log (marine)
Edward Massey *the younger, watchmaker of Hanley, Stoke-on-Trent, England.*

Tantalum
Anders Gustaf Ekeberg, *Swedish chemist and mineralogist. He chose the name because of the tantalising work involved in finding something to react with it.*

Ultra-violet rays
Johann Wilhelm Ritter, *scientist, Gotha, Germany, and separately,* William Hyde Wollaston, *philosophical writer, London.*

Wood-planing machine
Joseph Bramah, *London.*

1803
Atomic theory
John Dalton, *Quaker chemist of Manchester,* UK, *also a keen meteorologist.*

Cerium
Martin Heinrich Klaproth, *chemistry teacher, Berlin Artillery School, and named after Ceres, the asteroid – also discovered simultaneously by* Jöns Jakob Berzelius, *chemist, Stockholm School of Surgery, and* William Hisinger, *mining chemist, Sweden.*

Cinchonine
Andrew Duncan, *the younger, physician and professor at Edinburgh University,* UK.

Cotton-dressing machine
Thomas Johnson *of Bredbury, an ingenious but dissipated young weaver employed by Radcliffe & Ross of Stockport,* UK. *Johnson's machine, called the 'Dandy loom', enabled the warp to be dressed before it was put on the loom. Johnson received a bonus of £50 from the patentees, his employers, for his ingenuity.*

Fourdrinier paper-making machine
Bryan Donkin, *Bermondsey engineer, commissioned by John Gamble and financed by London stationers, Henry and Sealy Fourdrinier. The first*

machine, at *Two Water Mills,
Frogmore, Hertfordshire,* UK,
produced 600 lb of paper in 24 hours.

Iridium
Descotils *and* Smithson Tennant,
chemist, Cambridge, UK.

Spray gun
Dr Alan de Vilbiss, *medical
practitioner of Toledo, Ohio,* USA,
*while searching for the ideal method of
applying medication to oral and nasal
passages without the use of a swab.*

Three-furrow plough
Reverend Edmund Cartwright,
*whilst in charge of the Duke of
Bedford's experimental farm,* UK.

1804

**Bomb-proof round defence
tower (martello)**
Brigadier-General William Twiss,
*commanding engineer of the Southern
District of the* UK. *It was based on a
model of the defence tower at Mortella
Point, Corsica, brought back in 1794
by Admiral Jervis of* HMS *'Victory'.
Over 70 towers were built, but they
were never used for defence.*

Catamaran fire ship
Sir Sydney Smith, *in an unsuccessful
attempt to destroy the Boulogne flotilla
established by Napoleon to invade the*
UK.

**Double-cylinder expansion
steam engine**
Arthur Woolf, *mining engineer,* UK.

Elliptic carriage spring
Obadiah Elliott, *coachbuilder,
Lambeth, London.*

Embroidery machine
John Duncan, *weaver, Glasgow,* UK.

Fishnet-making machine
Joseph Marie Charles Jacquard,
*Lyons-born mechanic. Having seen the
machine, Emperor Napoleon asked,
'Are you the man who can do what God
Almighty cannot – tie a knot in a taut
string?' 'I can do not what God cannot,
but what God has taught me to do,'
replied Jacquard.*

Flinders bar
Captain Matthew Flinders, UK,
*following his observation regarding
compass deviation during a voyage to
Australia on the iron-hulled* HMS
*'Investigator'. Flinders's solution was
a vertical bar of soft iron.*

Gay-Lussac's Law
Louis-Joseph Gay-Lussac, *assistant
to Count Claude Louis Berthollet,
Arcueil, France.*

Osmium
Smithson Tennant, *chemist,
Cambridge,* UK.

Palladium
Dr William Hyde Wollaston,
chemist, UK.

Quinquet lamp
Aimé Argand, *Swiss physician and
chemist working in the* UK.

Steam locomotive (railway)
Richard Trevithick, *Cornish-born
steam-road-carriage engineer for the
Penydaren Iron Works, South Wales,*
UK. *Pulling five wagons, 10 tons of iron
and 70 men, it travelled at 5 mph on
the 10-mile track.*

Trevithick's locomotive

1805

Amphibious vehicle
Oliver Evans, *whose 'Orukter
Amphibolos' (amphibious digger) was
designed and built for the commercial
purpose of dredging the Philadelphia
Docks. A single cylinder steam engine
with a 20-inch boiler was geared to
drive either a wooden-wheeled wagon
or an eight-bladed stern paddlewheel.*

Chimney-sweeping machine
Smart.
Circular-blade mower for corn or grass
Thomas James Plucknett, *of Deptford, England.*
Gas-lit cotton mill
Phillips & Lee *of Manchester,* UK.
1000 separate burners facilitated night-shift work.
Mechanical silk loom
Joseph Marie Jacquard, *silk-weaver of Lyons, France. Based on weaving machines of* B Bouchon, H Falcon and Jacques Vaucanson *and also through study of Vaucanson's automata toys at the Paris Conservatorium. By 1834, some 30,000 Jacquard looms were at work in Lyons.*

Jacquard's punch-card loom

Morphine
Friedrich Wilhelm Adam Sertürner, *apothecary, Hameln, Germany.*
Rhodium
Dr William Hyde Wollaston, *chemist, London.*

War rocket (gunpowder in sheet-iron casings)
Sir William Congreve, *son of the Comptroller of the Woolwich Arsenal, London. In 1807, during the siege of Copenhagen, 40,000 such rockets were launched.*

1806
Bank-note-numbering machine
Joseph Bramah, *so enabling the Bank of England to dismiss 100 clerks.*
Beaufort wind scale
Rear-Admiral Sir Francis Beaufort, *hydrographer, as a set of numbers to indicate the strength of the wind from a Calm 0 to a Hurricane Force 12, 'that which no canvas could withstand'. It was accepted by the British Admiralty in 1838 and by the International Meteorological Committee in 1874.*
Carbon paper
Ralph Wedgwood *of London.*
Cast-iron printing press (rounded cheek frames)
Charles Mohun, *3rd Earl of Stanhope,* UK, *manufactured by Robert Walker.*
Coffee pot (built-in metal sieve)
Benjamin Thompson, Count Rumford, *Anglo-American administrator and scientist.*
Reflecting goniometer
Dr William Hyde Wollaston, *chemist, London.*

1807
Camera lucida
Dr William Hyde Wollaston, *chemist, London.*
Gas-lit street
Golden-lane, London, UK.
Patent for a gas-driven automobile
Isaac de Rivez.
Percussion-cap (arms)
Reverend Alexander John Forsyth, *former minister of Belhevie, Aberdeenshire,* UK. *He was pensioned by the British government when he refused to sell the secret to Napoleon.*

Potassium
Humphrey Davy, *Professor of Chemistry, Royal Institution, London.*
Sensory versus motor nerves (brain)
Charles Bell, *Scottish anatomist and surgeon.*
Sodium
Humphrey Davy, *Professor of Chemistry, Royal Institution, London.*
Steamboat (long-distance)
Robert Fulton, *former miniaturist and landscape painter, US. His 150 ft paddlewheeler, 'Clermont', travelled 150 miles on the Hudson River between New York and Albany at an average 5 mph.*
Windmill patent sail
William Cubitt, *miller, millwright and cabinet-maker of Norfolk, UK. It combined hinged shutters with remote control by chains from the ground. The patent sail and fantail were adopted in Denmark, Germany and the Netherlands.*

1808
All-iron plough
Robert Ransome, *engineer, of Ipswich, UK.*
Band-saw
William Newberry, UK. *It did not come into regular use until 1850 when better methods of steel-making and tempering removed the danger of the snapping of the band.*
Barium
Humphrey Davy, *Professor of Chemistry, Royal Institution, London.*
Bermuda sail (yacht)
Harvey, *a local resident of the Islands of Bermuda, who rigged his schooner with high, lateen-like sails, instead of the usual gaff rig.*
Boron
Joseph Louis Gay-Lussac, *Professor of Chemistry, École Polytechnique, Paris, and his associate* Louis Jacques Thénard.
Calcium
Humphrey Davy, *Professor of*

Chemistry, Royal Institution, London.
Circular railway
Richard Trevithick, UK *engineer near Euston Road, London. His locomotive 'Catch-Me-Who-Can' achieved 12 mph.*
Cotton-lace/bobbin-net machinery
John Heathcoat. *He set up a factory in Nottingham, UK, which was destroyed by the Luddites, forcing him to move business to Tiverton, Devon. Heathcoat also innovated ribbon- and net-making machinery.*
Electric-arc lamp
Humphrey Davy, *Professor of Chemistry, Royal Institution, London.*
Magnesium
Humphrey Davy, *Professor of Chemistry, Royal Institution, London.*
Steamboat (offshore)
John Stevens, US. *'Phoenix' steamed from New York out into the Atlantic around Cape May and up the Delaware Bay to Philadelphia.*
Typewriter (for the blind)
Pellegrino Turri *of Castelnuovo, Italy, for his blind friend, Contessa Carolina Fantoni da Fivizzano. It was a 27-charactered machine.*

1809
Heat-bottled food
Nicholas Appert, *chef and confectioner, Massey, France, following fifteen years' research. He was awarded 12,000 francs by the French Manufacturers' Bureau, provided he publish his process in detail. His pamphlet was translated and published into German, English and Swedish and was widely adopted. The House of Appert was formed in 1812.*
Lamarckism
Jean Baptiste Pierre Antoine de Monet, *Chevalier de Lamarck, naturalist and evolutionist, France.*
Paper-making machine (cylindrical)
John Dickinson *of Kings Langley,*

Hertfordshire, UK. *This type was developed for the manufacture of higher-grade papers.*

1810
Ammonia-soda reaction
Augustin Jean Fresnel, *French chemist.*
Electrolytic telegraph
S T von Sömmering, *Bavarian Academy of Sciences, Munich, at the command of Napoleon's ally, Margrave Leopold of Bavaria. This slow and complicated system was soon replaced by the electric telegraph.*

Sömmering's electro-chemical telegraph

Hemp and flax-spinning machine
Philippe de Girard.
Homeopathy
Dr Samuel Hahnemann of Leipzig, *who propounded his theories in the 'Organon of Medicine' and other works.*
Hydro-aeronaut (navigator's lifebuoy)
Thomas Cleghorn, UK.
Metronome
Dietrich Nikolaus Winkel, *German-born master organ-builder living in Amsterdam. Winkel demonstrated the instrument to Johann Nepomuk Maelzel, who patented it as his own invention and set up a factory to produce it as the Maelzel Metronome, despite an investigation in Paris which upheld Winkel as the true inventor.*

Maelzel's friend Beethoven was the first composer to include metronome markings on his scores in 1816.
Mowing machine
Peter Gaillard, *Lancaster, Pennsylvania*, USA.
Panomonico (or panharmonium)
Johann Nepomuk Maelzel *of Vienna, after seven years' development. 300 separate musical instruments were built into a one-man console. It was sold for 100,000 francs to the Archduke Charles of Austria for the exclusive purpose of annoying his friends. Beethoven wrote his 'Wellington's Victory at Vitoria' (1823) to be specifically played on this instrument.*

1811
Animal bone charcoal as a filter for sugar
Louis Figuier, *French chemist at the Paris School of Pharmacy.*
Avogadro's Law
Amedeo di Quaregna e di Ceretto, Count Avogadro, *scientist, Turin, Italy.*
Iodine
Bernard Courteois, *chemist and saltpetre manufacturer near Paris, France.*
Iron dome
Over the Paris Corn Market, France.
London Mint coin-milling machinery.

1812
Canned food
Bryan Donkin, *partner in John Hall's Dartford Iron Company*, UK. *He used iron cans coated with tin. The* UK's *first cannery was set up in Blue Anchor Road, Bermondsey. Among the first users of canned food – particularly preserved meat – were naval ships on longterm exploration voyages.*
Locomotive (rack and pinion gear)
John Blenkinsop, *Middleton Collieries, near Leeds*, UK.

Rotary printing press (steam-powered)
Friedrich König, *German engineer working in London, with chief mechanic,* André Friedrich Bauer, *and their financial backer,* Thomas Bensley. *The first commerical model was acquired by John Walker II for 'The Times' newspaper, London, in 1814. Its output was 1400 printed sheets per hour on one side only.*

Steamship (regular commercial service)
Henry Bell's *30-ton 'Comet' on the Clyde between Glasgow and Helensburgh,* UK.

Sugar-refining vacuum pan
the Hon. Edward Charles Howard.

1813

Air beds and water beds (india-rubber cloth)
John Clark, *grocer, Bridgewater, Somerset, England.*

Armoured warship (steam-powered)
Robert Fulton's *'Demologos'. It was 140 ft long with thirty 32-pounder cannon on board.*

Cartridge (practical)
Samuel Pauly, *Paris, for a .59 breech loader.*

Gas meter
Samuel Clegg, *chief engineer to the Chartered Gas Light and Coke Company in London, formed the previous year by an energetic German,* Friedrich Albert Winzer, *who had anglicised his name to* Winsor.

Liquid-filled compass
Francis Crow, *Faversham,* UK.

Teak-built ship
HMS *'Cornwallis', launched from the Bombay dockyard, India.*

Toxicology
Mathieu Joseph Bonaventure Orfila, *French chemist.*

1814
Planimeter
J M Hermann. *The first instrument was constructed to his designs in 1817.*

Plough (replaceable cast-iron parts)
John Jethro Wood, *Poplar Ridge, New York,* USA.

Railway gauge (4 ft 8½ in)
George Stephenson, *colliery engine-wright at Killingworth,* UK *for his locomotive 'My Lord', which attained a speed of 6 mph.*

1815
Alkalimeter
Andrew Ure, *Professor of Chemistry, Anderson's College, Glasgow,* UK.

'The Dennett Gig' (horse-drawn carriage)
Bennett, *coachbuilder of Finsbury, London. It was named after Miss Dennett, a popular dancer.*

Jaunting car (Irish)
Carlo Bianconi, *Milan, Italy. To drive between Clonmel and Caher, Tipperary, Ireland, each day, charging 2d a mile. By 1837 Bianconi*

Davy's miner's safety lamp (see overleaf)

had established 67 such conveyances, drawn by a total of 900 horses.

Miner's safety lamp
Independently by Sir Humphrey Davy, chemist for the Society for Preventing Accidents in Coal Mines, and by George Stephenson, colliery engine-wright. The former became known as 'The Miner's Friend', while the latter was nicknamed 'The Geordy'.

Pruning shears (sécateurs)
Bertrand, Marquis of Moleville, France.

1816

British Admiralty semaphore telegraph
Admiral Sir Home Riggs Popham. Although he produced the first edition of his Telegraphic Signals in 1800, the official trial between London and Chatham, with semaphores mounted on ships, took place almost twenty years later. Once proved, lines were established from Whitehall to Portsmouth, to Plymouth, to Chatham and to Dover. Popham received £2000 for his invention.

Circular saw (tempered steel teeth)
Auguste Brunet and Jean-Baptiste Cochot, France.

Compressed-air fire extinguisher
Captain George William Manby of London.

Copper percussion cap (arms)
Joshua Shaw, English portrait painter.

Electric telegraph
Francis Ronalds in his garden at Hammersmith, London. Although he published a description of it in 1823, nobody showed any interest.

'Heliographie' (sun-drawing)
Joseph Nicéphore Niépce, amateur scientist of Chalon-sur-Saône, France. He adapted the conventional 'camera obscura' to fix his images on a zinc or pewter plate, made light-sensitive by bitumen of Judea, a kind of asphalt normally used in engraving and here dissolved in oil of lavender. Niépce's exposure time was eight hours on a summer's day and his invention was the forerunner of photography.

Kaleidoscope
Dr David Brewster, Scottish physicist and Edinburgh magazine editor.

Oxyhydrogen blowpipe
Robert Hare, Professor of Chemistry, medical faculty, Pennsylvania University, USA.

Print roller
manufactured following a suggestion of William Nicholson, editor of the 'Philosophical Journal', UK.

Psychrometer
Joseph Louis Gay-Lussac, French chemist and physicist, for measuring atmospheric humidity.

Stethoscope (wood-turned cylinder)
René Theophile Hyacinthe Laënnec, pupil of Napoleon's personal physician, Corvisart, and chief physician to the Hôpital Necker, France.

Stirling-cycle engine
Reverend Robert Stirling, Dumbarton, Scotland, as a hot-air external combustion engine, with a regenerator that prevented dissipation of heat between cycles.

1817

Apollonicon
Benjamin Flight and Joseph Robson, organ-builders of Leicester Square, London. The most elaborate barrel-organ yet built, performing with either three barrels or six performers with one console each.

Dental plate
Anthony A Plantson, US.

Hobby-horse (steerable front wheel)
Baron Karl von Drais de Sauerbrun, Germany. It was called either 'draisina' or 'velocifère'.

Lithium
John August Arfwedson, *chemist and metallurgist, Sweden.*

Parkinson's Disease
James Parkinson, *surgeon and palaeontologist,* UK.

Pin-manufacturing machine
Seth Hunt, *Syracuse, New York,* USA. *Not in operation until 1824 when Samuel Wright took out a* UK *patent.*

Treadmill (prison)
William Cubitt *of Ipswich. The first such punishment machine was erected at Brixton prison, London.*

1818

Blood transfusion
Dr Thomas Blundell *of Guy's Hospital, London, who was unable to make his system work successfully through insufficient knowledge of the nature of blood.*

Cadmium
Friedrich Strohmeyer, *Inspector General of the Apothecaries, Hanover, Germany.*

Detector lock
Jeremiah Chubb, *ironmonger in Portsea, the dockland area of Portsmouth,* UK. *Despite being given any tool necessary, blank keys, duplicate lock, as well as the offer of a free pardon from the government and £100 from Mr Chubb, a lockpicker on board the Portsmouth prison ship was unable to pick the detector lock after three months' work.*

Encke's Comet
Johann Franz Encke, *astronomer, Seeberg Observatory, Germany, mathematically predicted the orbit of a faint comet and successfully predicted its return.*

Geothermal energy experiment
F de Larderel, *French engineer in Tuscany, Italy. Larderel Power Station went into action in 1904.*

Hydrogen peroxide
Baron Louis-Jacques Thénard, *Professor of Chemistry, University of Paris.*

Lens-polishing machine
Josef von Fraunhofer, *physicist and toolmaker, Utzschneider Optical Institute, near Munich, Germany.*

Profile lathe
Thomas Blanchard *of Middlebury, Connecticut,* USA. *The machine did the work of thirteen operators and allowed a reduction in wood-working prices.*

Selenium
Johan Jakob Berzelius, *chemist and Secretary to the Swedish Academy of Science.*

Strychnine
Pierre-Joseph Pelletier *and* Joseph Bienaimé Caventou *at the School of Chemistry, Paris.*

1819

Alternation of generations
Adelbert von Chamisso, *French-born biologist and lyric poet, in Berlin.*

Blocks of chocolate (factory produced)
François-Louis Cailler *at Vevey, Switzerland.*

Dental amalgam
Charles Bell.

Dioptric system
Augustin Jean Fresnel, *French physicist, chief of the Department of Public Works, Paris. This was an arrangement of lenses for refracting light in lighthouses. Fresnel also invented 'Fresnel's rhomb', a special prism for producing circularly polarised light.*

Macadamising
John Loudon Macadam, *a tacksman of the Kaims colliery, Muirkirk, Ayrshire,* UK – *as an improved method of road-making. In his early experiments with firmly embedded layers of granite chips, Macadam was working alongside his second cousin, Lord Dundonald.*

Magnetic field
Hans Christian Oersted, *Danish physicist, while giving an evening lecture on the galvanic battery at Kiel University. A wire from the battery*

accidentally fell on a mariner's compass, deflecting its needle from Polar North. Thus Oersted found the link between electricity and magnetism.
Naphthalene
John Kidd, *Aldrichian Professor of Chemistry at Oxford,* UK.
Open diving dress
Augustus Siebe, *Saxon engineer working in Soho, London.*
Siren
Baron Charles Cagniard de la Tour, *Parisian physicist, for measuring sound frequency.*
Transatlantic ship (sail with auxiliary steam)
Francis Fickett, *boatbuilder, and* Stephen Vail, *engine-wright. The 'Savannah' took 29½ days from Savannah, Georgia,* USA *to Liverpool,* UK.
Trigonometrical Ordnance Survey map of the UK
Printers as directed by Colonel Mudge, *following commencement of the survey back in 1783.*

1820
Arithmometer
C X Thomas *of Colmar, Alsace. Sold in large numbers 40 years later to assurance companies.*
Corn cultivator
Henry Burden, *of Troy, New York,* USA.
Diphtheria
Pierre Fidèle Bretonneau, *physician and bacteriologist, Tours, France.*
Gas-lit city
London; Paris at the same time; Dublin by 1825; Sydney, Australia by 1841.
Hygrometer (ether)
John Frederic Daniell, *chemistry professor at King's College, London.*
Medical stitching wire
Pierre-François Percy, *French surgeon.*
Quinine
Pierre Joseph Pelletier *and* Joseph Bienaimé Caventou *at the School of Chemistry, Paris.*

Soda-water-making apparatus
Charles Cameron, UK.
Tilbury gig (carriage)
Thomas Tilbury, *coachbuilder of Marylebone Road, London for the Hon. Fitzroy Stanhope.*

1821
Anchor (stockless patent)
Richard Francis Hawkins, *Master Mariner, Plumstead, Kent,* UK.
Caffeine
Pierre Joseph Pelletier *at the School of Chemistry, Paris.*
Dark lines (solar spectrum)
Josef von Fraunhofer, *Director of the Physical Cabinet, Munich, Germany.*
Eidograh
William Wallace, *Professor of Mathematics, Edinburgh University,* UK.
Enchanted lyre
Charles Wheatstone, *musical instrument-maker, London. It conveyed the sounds of a musical box from a cellar to upper rooms by means of a deal rod.*
Heliotrope
Carl Friedrich Gauss, *German mathematician, astronomer and physicist, to conduct a trigonometrical survey of Hanover.*
Mouth organ
Friedrich L Buschmann *of Berlin, as a free-reed musical instrument he called the 'Mundäoline'.*
Multiplying galvanometer
Johann Christian Poggendorff, *German physicist and chemist, aged 25.*

1822
Accordion
Friedrich L Buschmann *of Berlin, as a bellow-vibrated piano keyboard. He called it the 'Handëoline'.*
Aquatic tripod
William Kent *of Lincolnshire,* UK. *It was used to sit on and paddle while duckshooting.*

Buschmann's 'Handëoline' or accordion

Bauxite
Pierre Berthier, *French mineralogist, near Baux, Provence, France.*
Diorama
Bouton *and* Louis Jacques Mande Daguerre, *scene painter in Paris.*
Thermocouple
Thomas Johann Seebeck, *physicist, Berlin Academy. He found that when one pair of dissimilar wires with intertwined ends is placed in a hot surround, and another pair in a cold surround, an electric current flows.*
'Volontas' (horse-drawn carriage)
Miln Parker, US *for Cuba and Mexico.*

1823
Dowed zither
Johann Petzmayer, *Munich, Bavaria.*
Döbereiner lamp
Johann Wolfgang Döbereiner, *German chemist in Jena.*
Double-escapement piano
Sébastien Érard *of Paris.*
Hackney cabriolets
David Davies, *coachbuilder of Albany Street, London.*
Meteorological kite
Reverend George Fisher *at Igloolik, while on Captain William Parry's second Arctic exploration voyage. He attached a Six thermometer to a paper kite.*
Omnibuses
Stanislaus Bawdry, *Proprietor of the*

hot water baths in a suburb of Nantes, France.
Robinson's Patent Barley Water
Matthias Archibald Robinson, *grocer of St George the Martyr, London.*
Roller-skates (five wheeled)
Robert Tyers, *London fruiterer.*
Silicon
John Jakob Berzelius, *Swedish chemist.*
Talking doll
Johann Nepomuk Maelzel. *She said 'Maman' and 'Papa' and cost 10 francs.*
Waterproof rubber
Charles Macintosh, *manufacturing chemist, Glasgow, UK, using the rubber-dissolving naphtha by-product from the recently established Glasgow Gasworks. He went into partnership with H H Birley to build a factory for the manufacture of waterproofed capes and coats.*

1824
Artificial cement
Joseph Aspdin, UK. *Called 'Portland cement', it was further perfected in the 1840s by Aspdin's son William, and by Isaac Thompson, London builder.*
Electromagnet
William Sturgeon, *lecturer in Science, Royal Military Academy, London. He found that any piece of soft iron could be turned into a temporary magnet by putting it in the centre of a coil of insulated wire and making electric current flow through the coil (see overleaf).*
Galvanometer
André-Marie Ampère *at the National Institute of Paris, to measure the flow of electricity with a free-moving needle. The amp is named after him.*
High-speed self-acting mule
Richard Roberts, *inventor, Manchester, UK.*
Magnetic pull
François Jean Dominique Arago, *Parisian astronomer, whose non-*

Sturgeon's electro-magnet (previous page)

magnetic disc could bring a vibrating
magnetic needle to rest and could then,
mysteriously, pull the needle round
with it when rotated – but without
contact.

Plaited candle wick
Jean Jacques Cambacères, *Paris,
France.*
**Steam carriage (four-wheel-
drive)**
Timothy Burstall *and* James Hill.
Steam gun
Angier March Perkins, UK. *However,
this proved ineffectual.*
Veneer-cutting machine
Alexander Craig, UK.
Wine-pressing machine
Ignazio Lomeni, *Italy.*
Wrought-iron I-shaped beam
Thomas Tredgold, *engineer,* UK. *It
was not used practically until 1841 by
Robert Stephenson for railway bridge
construction.*

1825
Actinometer
Sir John Herschel, UK *astronomer, to
measure the heating power of solar
rays.*
**Binocular telescope (production
model)**
J P Lemière, *Paris.*
Concertina
Professor Charles Wheatstone,
London.
Double-action pedal-harp
Sébastien Érard, *piano-maker, Paris.*
Frigate (iron-masted)
HMS *'Phaeton'.*
Regular locomotive route
George Stephenson, *consulting
engineer for the Stockton and
Darlington mineral railway, near
Newcastle,* UK. *His 'steam blast' device
gave both his locomotives a practical
investment.*
**Rubber driving belts
(industrial)**
Thomas Hancock, UK.
Thaumatrope (optical toy)
Dr John Ayrton Paris, UK.

1826
Aniline (distilled indigo)
Otto Unverdorben, *chemist,
Potsdam, Germany.*

Biela's Comet
Wilhelm von Biela, *Austrian army officer and amateur astronomer.*
Bromine
Antoine-Jérôme Balard, *chemistry demonstrator, School of Pharmacy, Montpellier, France.*
Drummond Light
Lieutenant Thomas Drummond, *following a design by* Goldsworthy Gurney. *In 1820 Drummond had joined the Ordnance Survey project and used his 'limelight' and his improved heliostat for the work in hand. At night the Drummond Light could be seen at a distance of 112 miles.*
Friction match
John Walker, *chemist of Stockton-on-Tees, County Durham,* UK. *Originally made of cardboard coated with potash and antimony, packed by the 100 in tin tubes, and ignited using glass paper, they were nicknamed 'Lucifers'.*
Gas stove
James Sharp, *assistant manager of the Northampton Gas Co.,* UK. *He installed it in the kitchen of his Northampton home.*
Illuminated clock dial
St Bride's Church, *Fleet Street, London; lit by twelve gas-burners.*
Kite-carriage
Colonel Viney *and* George Pocock, UK. *Using one or more kites they made a journey from Bristol to London. The following year, on the Reading to Windsor road, the galloping horses pulling the Duke of Gloucester's carriage were unable to keep pace with the kite-carriage.*
Tea in packets
John Horniman *of Ryde, Isle of Wight,* UK.
Velocity of sound (in water)
Jacques Charles François Sturm, *Swiss mathematician, by means of a bell submerged in Lake Geneva.*

1827
Aluminium
Friedrich Wöhler, *chemistry teacher,*

Berlin, Germany.
Artificial synthesis of the urea molecule
Friedrich Wöhler, *as above.*
Astigmatic lens
George Biddell Airy, *Lucasian Professor of Mathematics, Cambridge,* UK. *It was manufactured the same year by Messrs Fuller of Ipswich.*
Electrical resistance
George Simon Ohm, *while professor at the Jesuit College in Cologne. On publication of his discovery, Ohm was so disappointed at its poor reception that he resigned his post. His name now represents the unit of electrical resistance.*
Floorboard-planing machine
Malcolm Muir *with* William Thomson, UK.
Mechanically-produced pasta factory
Buitoni *of Sansepolcro, Italy.*
Microphone
Professor Charles Wheatstone, *London.*
Microscope (dioptric-achromatic)
Giovanni Battista Amici, *Director of the Observatory, Royal Museum, Florence, Italy.*
Shorthand-writing machine
Gonod, *librarian at Clermont-Ferrand, France.*
Trifocal lenses
John Isaac Hawkins.
Water turbine
Bénôit Fourneyron, *student of Professor Claude Burdine, St Etienne School of Mines, France. By 1833 Fourneyron had developed a 50 hp turbine spinning at an unheard-of 2300 rpm.*

1828
Artificial ultramarine blue (pigment)
simultaneously by Guimet *and* Leopold Gmelin, *Professor of Chemistry, Heidelberg, Germany.*

Beryllium
Friedrich Wöhler, *chemistry teacher, Berlin, Germany.*

Chairs on small carriage springs
Samuel Pratt, UK.

Cocoa
Coenrad van Houten *of Amsterdam, by extracting excess cocoa butter from the crushed cacao bean.*

Corrugated iron
Richard Walker, *builder, of Rotherhithe, London.*

Differential gear
Onésiphore Pecqueur, *France, for steam-carriages.*

Hot-blast iron smelting
James Beaumont Neilson, *foreman and manager of Glasgow Gasworks,* and Charles Macintosh.

Reaping machine
Patrick Bell, *divinity student and son of a farmer in Angus,* UK, *together with the local blacksmith. By 1832 only ten Bell reapers had between them cut 400 acres of corn in Scotland.*

Ring-spinning frame
J Thorpe, UK.

Stethoscope (with earpiece)
Pierre Adolphe Poirry, *France.*

Thorium
Johan Jakob Berzelius, *chemist and Secretary of the Swedish Academy of Science.*

1829

Aeolina
Professor Charles Wheatstone, *as a free-reed hexagon-shaped musical instrument.*

Electro-magnetic motor
Joseph Henry, *Professor of Mathematics and Natural Science at the Albany Academy, New York,* USA. *Henry used a powerful short coil magnet for his innovation.*

Filtration plant for water
James Simpson, *Chelsea Water Works, London.*

The Graham cracker (biscuit)
Sylvester Graham, US *food doyen.*

'The most improved locomotive engine'
George Stephenson's *Newcastle-built 'Rocket', beating both the 'Novelty' and 'Sans Pareil' locomotives at Rainhill, near Liverpool,* UK, *by averaging 15 mph for the total of 60 miles, hitting a top speed of nearly 30 mph on the last lap.*

Steam-powered fire engine (horse-drawn)
John Braithwaite *and* John Ericsson *of London. Although it was successful at the Argyll Rooms, the English Opera House and Barclay's brewery, orders for the engine were not forthcoming in the* UK, *although Braithwaite sold an improved version to the King of Prussia for use in Berlin.*

Steam-jet-blast coach
Sir Goldsworthy Gurney *of London. Two years later, Charles Dance ran a steam-coach service between Gloucester and Cheltenham (four round trips daily). Before it was suspended, his fleet had travelled a total 4000 miles and carried 3000 passengers.*

Swinging-sector writing machine
William Austin Burt, *Michigan,* USA. *Called 'the Typographer'.*

'True water cure'
Vincent Priessnitz, *Grafenberg, Silesia.*

1830

Bi-metal thermostat
Andrew Ure, *Professor of Chemistry, Anderson's College, Glasgow,* UK.

Elastic web
Messrs Rattier *and* Guibal *at their waterproof cloth factory in Saint-Denis, Paris, assisted by machinery and workmen supplied by Thomas Hancock* UK *two years earlier.*

Lawn mower
Edwin Beard Budding, *textile engineer, Gloucester,* UK, *in partnership with* John Ferrabee. *'Country gentlemen may find in using*

Budding's patent grass-cutting machine

my machine themselves an amusing, useful and healthy exercise.'
Net-weaving machine (variable pitch mesh)
Alexander Buchan, *Scottish engineer.*
Paraffin
Karl, Baron von Reichenbach, *German natural philosopher and industrialist of Blansko, Moravia.*
Phosphorous matches
Charles Sauria, *student at Dole College in the Jura Mountains, France. They were manufactured commercially by Frederick Kammerer in Germany, and known as 'Congreves'.*
Pig-iron (boiling)
Joseph Hall *of Tipton, Staffordshire,* UK.
Sewing machine
Barthélemy Thimmonier, *poor tailor of St Etienne, France, chiefly for the purpose of making army clothing. Although he took out patents in both the* UK *and* USA *in 1848, Thimmonier was unable to market his invention and died in poverty in 1857.*

Tubular boiler
James Napier, *marine engineer, Glasgow,* UK.
Vanadium
Nils Gabriel Sefström, *Swedish physician and chemistry teacher, School of Mines, Stockholm, Sweden.*

1831
Alizarine red
Robiquet *and* Colin, *as a crystalline body, the colouring principle of madder. It was synthesised by* Graebe and Liebermann *in 1869.*
'Bogie' truck (railroads)
John Jervis, *locomotive builder of New York,* USA.
Electric bell
Joseph Henry, *mathematics teacher at Albany, New York,* USA.
Electro-magnetic balance
Antoine César Becquerel, *Paris.*
Electro-magnetic induction
Michael Faraday FRS, *Fullerian Professor of Chemistry at the Royal Institution, London. He also discovered the principle of the*

transformer.

Magnetic North Pole
Commander James Clark Ross,
during Sir John Ross's second voyage.

Reaping machine
Cyrus McCormick, *22-year-old son
of Robert McCormick, farmer and
inventor of Virginia,* USA. *The
McCormicks sold their first reaper as
late as 1840 and by 1844 50 had been
sold. Soon after, they moved to
Chicago and set up a factory which by
1871 was turning out 10,000
harvesters a year.*

Steam-powered omnibus
Walter Hancock, UK. *Called 'The
Infant', it ran between London and
Stratford.*

Thermo-multiplier
Macedonio Melloni *and* Leopoldo
Nobili, *Italian physicists.*

Waistcoat-pocket pedometer
William Payne, *watch and
clockmaker, London.*

1832
Alcohol still
Aeneas Coffey, *distiller, the Dock
Distillery, Dublin.*

**Artist's watercolour cake
(glycerine-based)**
William Winsor *and* Henry Charles
Newton, *artists' colourmen of
Rathbone Place, London.*

Collapsible silk opera hat
Antoine Gibus, *Paris hatter.*

Hydraulic-powered factory
E Egberts *and* Timothy Bail, US.

Hydrostatic bed
Neil Arnott, *Scottish physician to the
Spanish Embassy, London.*

Kindergarten
Frederick Wilhelm August Froebel,
*former science teacher at Blankenburg,
Switzerland.*

Magneto-electric machine
Hippolyte Pixii, *Paris, and in 1833 by*
Joseph Paxton, *Cambridge,* UK.

**Motion-picture machine
(primitive)**
simultaneously by Dr Joseph Antoine

Ferdinand Plateau (*Phenakistoscope*)
in Ghent, Belgium, and Simon Ritter
von Stampfer *(Stroboscopic disc) in
Vienna, Austria.*

**Multiple-effect sugar
evaporation**
Rillieux.

**Passenger-tramway (horse-
drawn)**
New York to Harlem, USA.

**'Wheel of Life/Daedalum'
(optical toy)**
W H Horn *of Bristol,* UK. *Patented in
the* USA *in 1867 by* Milton Bradley.

1833
Book dust-jacket
Messrs Longman *of London for their
annual 'The Keepsake'.*

Creosote
Karl Baron von Reichenbach,
*German natural philosopher and
industrialist of Blansko, Moravia.*

Cycloidal paddle-wheel
Joshua Field, *civil engineer, Lambeth,
London.*

Diastase
Anselm Payen, *Professor of Industrial
Chemistry in Paris, and* J F Persoz,
*Professor of the School of Pharmacy at
Strasbourg.*

Differential calculating machine
Charles Babbage, *Lucasian Professor
of Mathematics at Cambridge,
working in London. Babbage began
work in 1821 on a machine that would
prevent errors occurring in
mathematical and astronomical tables.
During the next decade the government
subsidised him to the tune of £15,000
and would have continued, but in 1834
Babbage had yet another argument
with his skilled instrument-maker, who
resigned. The portion completed was
placed in the library of King's College,
London.*

Electric telegraph key
Carl Friedrich Gauss, *Professor of
Mathematics, and* Wilhelm Eduard
Weber, *Professor of Physics,
Göttingen University, Germany.*

Hansom's patent safety cab, 1834

Fluorescence in chlorophyll
Sir David Brewster, *Scottish chemist.*
Glacial fossil fishes
Jean Louis Rodolphe Agassiz, *Swiss naturalist.*
Kryptographic pen
Xavier Progrin *of Marseilles, France.*
Lock-stitch sewing machine
Walter Hunt, US.
Nervous reflex
Marshall Hall, *physician and physiologist, London.*
Nitro cellulose
Théophile-Jules Pelouze *and* Henri Braconnot, *French chemists.*
Oakey glasspaper
John Oakey *of North London. A piano-maker's son, some 30 years before he had experimented with gluing powdered glass on to old ledger sheets. For his trademark, Oakey chose the profile of the then popular Duke of Wellington.*

1834
Adhesive postage stamp
James Chalmers, *bookseller and newspaper publisher, Dundee,* UK.
Chinese white (watercolour pigment)
Messrs Winsor & Newton, *artists' colourmen of Rathbone Place, London.*
Continuous electric light

(galvanic cells)
James Bowman Lindsay, *self-taught scientist, Dundee,* UK.
Free-standing kitchen range
Philo Penfield Stewart, *missionary and teacher in Ohio,* USA. *It was called the 'Oberlin stove'.*
Hansom cab (carriage)
Joseph Aloysius Hansom, *inventor and architect, as his Patent Safety Cab. Substantial improvements were made by* John Chapman *in 1836.*
Machine-made tiles
Joseph *and* Xavier Gilardoni *at their factory in Altkirch, Alsace, France.*
Nocturne (musical)
John Field, *Irish composer.*
Pantascopic spectacles
George Richard Elkington, *optician and gilt toy-maker of Birmingham,* UK. *They were half-eye spectacles with joint angles.*
Powered roller-conveyor
Mr Grant *of the British Navy's Deptford Victualling Department, for fully mechanised 'hardtack' biscuit-baking.*
Threshing/fanning machine
John A *and* Hiram Abial Pitts, US.
Tonometer
H Schreiber, *Crefeld, for tuning musical instruments. It incorporated 52 tuning forks.*

Valve gear (marine steam engine)
Samuel Seaward, *London.*

1835
Automatic revolving cylinder gun
Samuel Colt, *former merchant seaman and chemistry lecturer of Hartford, Connecticut,* USA. *In 1846, for the Mexican War, the* US *government ordered 1000 Colt 'revolvers' for army use.*

Colt's revolver

Tonic sol-fa
Sarah Glover, *musical educationalist of Norwich,* UK*: 'doh, re, mi, fa, so, la ti, doh'.*
Tunnelling shield (cast-iron)
Marc Isambard Brunel *for the completion of the abandoned Rotherhithe-Wapping tunnel under the River Thames,* UK. *It was completed by 1843.*
Wallpaper (cylinder-printed)
Bumstead, UK. *The machine was hand-cranked and could print 200 rolls of wallpaper per day. Introduced to France by Isidore Leroy.*
Whit classification (standardised nuts and bolts)
Joseph Whitworth, *engineer, Manchester,* UK.

1836
Acetylene
Edmund Davy, *Professor of Chemistry, Dublin, Ireland.*
Bullet (conoidal cup rifle)
Captain Claude Etienne Minié *and* Captain Delvigne *of Vincennes, France. The first long bullet to replace the round metal ball.*

Colour printing
George Baxter. *In some of his illustrations to his 'Pictorial Album', he employed twenty different blocks.*
Combine harvester
H Hoare *and* J Hascall *of Michigan,* USA. *Although it initially failed because of climatic conditions, later it was taken to California and used successfully in 1854*
Daniell Cell
John Frederick Daniell, *first Professor of Chemistry at the newly founded King's College, London. Daniell's non-polarising design gave a more constant flow of electric current.*
Filing/shaping machine
James Nasmyth, *Scottish engineer, at his Bridgewater foundry, Patricroft, near Manchester,* UK.
Horse-box (purpose-built)
Messrs Herring, *coachbuilders of Long Acre, London, as commissioned by Lord George Bentinck, race-horse owner.*
Long-haired-silk-spinning machine
Gibson *and* Campbell *of Glasgow,* UK.
Standard test for arsenic (forensic analysis)
James Marsh, *chemist, Royal Military Academy, Woolwich,* UK.
Steam ram (for naval warfare)
James Nasmyth, *Scottish engineer.*
Steam shovel
William Smith Otis, *to dig the railbed of the Western Railroad, Massachusetts,* USA. *Soon after, it was used on railway projects in Maryland and Canada.*
Stroboscope
Joseph Antoine Ferdinand Plateau, *Professor of Physics at Ghent.*
Yeast growth during fermentation
Baron Charles Cagniard de la Tour, *Parisian physicist.*

1837
Axial turbine
Fontaine-Baron, *France.*

Bolt-action rifle ('needle gun')
Johann von Dreyse *of Sommerada,*
Prussia.

Braille
Louis Braille, *blind professor at the*
Institute for Blind Youth, Paris. The
43-symbol system was based on
artillery captain Charles Barbier's
dot-dash punching machine for battle-
field communication by night.

Closed diving dress
Augustus Siebe, *Saxon engineer*
working in London.

Daguerreotype (photographic
process)
Louis Jacques Mandé Daguerre,
former Parisian scene-painter, with
Claude de St V M F Niépce. Exposure
time was reduced to 20–30 minutes by
using mercury vapour and fixing with
a solution of common salt.

Electric motor
Thomas Davenport *of Rutland,*
Vermont, USA, for drilling holes and
turning hardboard. Three years later,
this motor powered the first electric
printing press.

Electric telegraph
William Fothergill Cooke *and*
Professor Charles Wheatstone *of*
King's College, London. In 1846
Cooke established the Electric
Telegraph Company and after six
years some 4000 miles of telegraph wire
had been installed.

Galvanised iron
William Henry Crauford, *London.*

Propeller (practical marine)
simultaneously by Francis Pettit
Smith, *Middlesex farmer, with the*
three-masted schooner 'Archimedes',
and Jon Ericsson, *ex-Swedish Army*
officer with the 10-knot 'Francis B
Ogden' and the 13-knot 'Robert F
Stockton' – both in the UK.

Shorthand system (phonetic)
Isaac Pitman, *aged 24,* UK.

Steam tram
New York and Harlem Railway, USA.

Steel plough
John Deere, *blacksmith and plough*
maker of Illinois, USA. The entire
plough was steel except its braces,
beams and handles.

Worcester sauce
John Lea *and* William Perrins,
pharmacists of Worcester, UK, to a
recipe of Sir Marcus Sandys.

1838

Brougham (carriage)
Named after Henry, 1st Baron
Brougham and Vaux, UK *politician*
and barrister, and allegedly built to his
own design, either by Robinson &
Cook of Mount Street, London or
Sharp & Bland of South Audley
Street, London.

Children's hosiery factory
(steam-powered)
John Button, UK *emigrant at*
Germantown, USA.

Cipher-writing machine
Antoine Dujardin, *Lille, France. It*
was called the 'Tachygraphe'.

Equirotal phaeton (horse-drawn
carriage)
W Bridges Adams, *sponsored by the*
Duke of Wellington.

Fluorescence in fluorspar
David Brewster, *Scottish physicist.*

Morse Code
Samuel F B Morse, *art professor, and*
his student Alfred Vail *at New York*
City University, USA. It was not until
1843 that the Morse Bill for the first
telegraph line from Washington to
Baltimore was passed in Congress by a
very narrow margin (see overleaf).

Paddleboat (electric)
Professor Moritz Hermann Jacobi,
German physicist and engineer at St
Petersburg, Russia. It attained 2 mph
along the River Neva.

Photographic prints (on silver
chloride)
William Henry Fox Talbot *of Lacock*
Abbey, Wiltshire, UK. Because the
Talbotype or Calotype involved a
positive-negative process allowing for
more than one print, it had soon ousted
the direct positive, unique-print

1838

Morse telegraph recorder

Daguerreotype. A similar system was evolved simultaneously by Hippolyte Bayard, *civil servant, France.*
Plant cells
Matthias Jakob Schleiden, *Professor of Botany, University of Jena, Germany.*
Secondary electric currents
Joseph Henry, US.
Steam hammer
James Nasmyth, *Scottish engineer, for forging an enormous wrought-iron paddleshaft that was never used. In 1842 he found his steam hammer at work in Le Creusot, France. It had been adapted from his own scheme-book. Nasmyth patented his invention and it was adopted by the Admiralty in 1843.*
Stereoscope (3-D pictures)
Charles Wheatstone, *Professor of Experimental Physics, King's College, London.*

1839
Animal cells
Theodor Schwann, *German physicist and Professor of Anatomy at the Roman Catholic University, Louvain, France.*

Artificial fertiliser
John Bennet Lawes, *for the cultivation of turnips at Rothamsted, Harpenden,* UK.
Babbitt Metal
Isaac Babbitt, *inventor, Boston, Massachusetts,* USA.
Brick-making machine
Messrs Cooke & Cunningham, *whose machine could make 18,000 bricks in ten hours.*
Chenille weaving process
James Templeton, UK.
Electrotype
Professor Moritz-Hermann Jacobi, *physicist and engineer at St Petersburg, Russia.*
Gas cooker
At Leamington Spa, UK, *where a dinner was cooked for 100 guests.*
Harrow (iron)
William George Armstrong, *inventor and solicitor, Newcastle,* UK.
'Jet-propelled' ship
J *and* M V Rutheven. *First tests were held on the Scottish Union Canal, near Edinburgh.*
Lanthanum
Carl Gustav Mosander, *Professor of Chemistry and Mineralogy,*

Stockholm, Sweden.
Manganese steel (crude)
Thomas Heath.
Odontology
Professor Richard Owen, *Professor of*
Comparative Anatomy at the College
of Surgeons, London.
Pedal bicycle (prototype)
Kirkpatrick MacMillan, *blacksmith,*
of Courthill, Dumfriesshire, Scotland.
A one-off, its furthest distance
travelled being 40 miles.
'Protoplasm' (the term or word)
Jan Evangelista Purkinje, *Professor of*
Physiology, Prague, Czechoslovakia.
Silurian system
Roderick Impey Murchison,
Scottish geologist.
Tenor-bass trombone
C F Sattler, *Leipzig.*

1840
Astro-photography
Dr John William Draper, *New York,*
USA. *He took a primitive*
daguerreotype of the moon. Three
years later he took a daguerreotype of
the solar spectrum.
Atmospheric railway
(pneumatic)
Samuel Clegg the Younger *and the*
Samuda brothers, Joseph and Jacob.
A line of rail was laid across
Wormwood Scrubs, London, between
Shepherd's Bush and the Great
Western Railroad, to test the efficiency
of atmospheric tubes, air pump and the
speed of carriages.
Box-girder construction
Andrew Thomson, *for a road bridge*
across the Pollock and Govan
Railway, Glasgow, UK.
Chronoscope
Professor Charles Wheatstone – *to*
measure small intervals of time.
Clarence carriage (horse-
drawn)
Named after the Duke of Clarence.
Collapsible metal tube (squeeze-
and-fold)
John Rand, US. *The following year,*

Messrs Winsor & Newton, *artists'*
colourmen of London, adapted the
process for their pigments.
Electroplating (silver-gold etc.)
John Wright. *Commercialised by the*
Elkington *cousins,* George Richard
and Henry, *cutlers of Birmingham,*
replacing Sheffield plate.
Ozone
Christian Friedrich Schönbein,
Professor of Chemistry at Basle, whilst
experimenting with William Grove's
battery.
Photographic lens (f/3.4)
Joseph Petzval, *Hungarian optician*
working for Voigtländer in Vienna.
Pimm's Number One (drink)
James Pimm, *London restaurateur.*
Prefabricated glass and iron
construction
Joseph Paxton, *gardener-architect,*
and Decimus Burton. *They built the*
Royal Conservatory at the Duke of
Devonshire's Chatsworth Estate, UK.
By 1850 Paxton had developed his
system for the Crystal Palace, London,
using 300,000 panes of glass, erected in
only eight months.
Pre-paid postage at low and
uniform rate
Rowland Hill, UK. *Called the 'Penny*
Post'.
Silvered glass process (mirrors)
Justus von Liebig, *Professor of*
Chemistry at Giessen.
Tonic water
Schweppes Co. *of England. Sugar*
and quinine were added to the soda
water recipe developed half a century
before by Jakob Schweppe, *German*
amateur chemist working in London.
Type-composing machine
James Young, UK, *and* Adrien
Delcambre *of Lille, France. Called*
the 'Pianotype', it was used in 1841 to
set a weekly magazine, 'The Phalanx',
published in London.
Xylophone (modern orchestral)
Zinc-coated ('galvanized') iron
sheet.

1841
Clod-breaking roller
William Croskill, UK.
Didymium
Carl Gustaf Mosander, *chemist,*
Stockholm.
Disposable book
Christian Bernhard Tauchnitz *of*
Leipzig with the 'Collection of English
Authors' series.
Street-cleaning machine
Joseph Whitworth, *mechanical*
engineer. It was introduced by the
Road and Street Cleaning Co. at
Manchester, UK.
Vulcanised rubber
Charles Goodyear, *former hardware*
merchant, Philadelphia, USA, *after a*
decade of struggle, poverty and
ridicule.

1842
Ether anaesthesia (surgery)
Crawford Williamson Long *of*
Jefferson, Georgia, USA, *physician,*
who first used it while operating on the
neck tumour of Mr James Venable, but
did not reveal his discovery until 1849.
Friction match-making
machine
Reuben Partridge.
Galvanic battery (carbon
electrodes)
Robert Wilhelm Bunsen, *German*
physicist and Professor of Chemistry at
Marburg.
Hyposulphite of soda
(photographic fixative)
John F W Herschel, UK *astronomer.*
Steam-powered planing
machine
Joseph Whitworth, *engineer, in*
Manchester, UK.
Submarine cable
Samuel Morse, US. *It was laid in New*
York harbour between the Battery and
Governor's Island.
Superphosphate (artificial
fertiliser)
Sir John Bennet Lawes *at*
Rothamsted, UK. *There were further*

experiments with J G Gilbert
throughout the 1840s.
Torpedo (static mine)
Samuel Colt, US *inventor. It was*
exploded underwater in New York
harbour by means of a galvanic battery
and later blew up a 500-ton ship on the
River Potomac, with Colt detonating
the circuit from 5 miles' distance.
Wagonette
Mr Lovell, *coachbuilder, of*
Amersham, UK, *for Lord Curzon.*

1843
Aneroid barometer (spring-
type)
Lucien Vidi, *France.*
Calculating machine
Pehr Georg Scheutz and his son
Edward, *engineers, of Stockholm,*
following a study of the Babbage
machine. Their machine was
purchased by J F Rathbone, US
merchant, to be presented to Dudley
Observatory, Albany, New York.
Christmas card
John Calcott Horsley RA, *on the*
suggestion of Henry Cole. 1000 were
printed by Messrs Jobbins of Warwick
Court, Holborn, London, and
individually hand-coloured.
Cigarettes (commercially-
produced)
Manufacture Français des Tabacs,
France.
Derringer (pocket pistol)
Henry Derringer, *small-arms*
manufacturer, Philadelphia, USA.
Erbium
Carl Gustaf Mosander, *chemist and*
mineralogist, Stockholm.
Gutta-percha
made known in Europe by Dr
Montgomery *and* Dr de Almeida,
who brought it back from Malaya some
200 years after John Tradescant,
English naturalist and gardener, had
brought small specimens to England.
Iron steamship (screw-
propelled)
Isambard Kingdom Brunel. *His* SS

'*Great Britain*' *made the crossing from Liverpool to New York in* 14½ *days.*

Joule
James Prescott Joule, *brewer and self-taught physicist, Manchester,* UK: *an energy unit to describe his first law of thermodynamics.*

Pantelegraph
Alexander Bain, *London.*

Photographic enlarger
Alexander Wolcott *and* John Johnson, US.

Quaternions (mathematics)
Sir William Rowan Hamilton, *mathematician and President of the Royal Irish Academy.*

Ridley-Bull wheat stripper
simultaneously and independently by John Ridley *and* John Wrathall Bull, *Southern Australia.*

Terbium
Carl Gustaf Mosander, *chemist and mineralogist, Stockholm.*

Textile mill (water-turbine-powered)
Uriah A Boyden, *New England,* USA.

Wood pulp (mechanically ground)
Friedrich G Keller, *weaver, of Haynich, Saxony. Keller sold his secret process to Heinrich Voelter for a pittance and the latter became rich by perfecting the process commercially.*

1844
Adhesive envelopes
to replace sealing wax.

Adjustable pitch propeller
Bennet Woodcroft, *consulting engineer and patent agent.*

'Aerial wheels' (pneumatic and bolted on)
Robert William Thompson, *civil engineer, of London. Of india-rubber or gutta-percha coated on to canvas and filled with air, they were hand-made, hence expensive and not easily replaceable.*

Boat-cloak
Lieutenant Peter Alexander Halkett, *Royal Navy, as a garment which could be turned into an inflatable boat. Halkett's many demonstrations of the boat-cloak culminated in the Bay of Biscay where he successfully paddled with a convertible umbrella-sail from* HMS '*Caledonia*' *to* HMS '*Vincent*' *of the Experimental Squadron.*

Cigarette-manufacturing machine
Le Maire, *France.*

Crèches
established in Paris, to receive the young children of working mothers, they came to London in 1863.

Napier's coin-weighing machine
for use in the Bank of England.

Nitrous oxide anaesthetic
Horace Wells, *dentist, and* Gardner Quincy Colton, *chemist, at Hartford, Connecticut,* USA.

Paper-cutting machine
Guillaume Massicot, *France.*

Ruthenium
Karl Karlovitch Klaus, *chemistry lecturer, Kazan University, Russia.*

Saxophone
Antoine Joseph Sax, *instrument-maker of Belgium. In* 1857 *Sax was appointed saxophone instructor at the Paris Conservatoire (see overleaf).*

Spinning rocket (angled exhaust outlets)
William Hale, UK.

1845
Ball-valve (reliable)
Edward Chrimes *of Rotherham, Yorkshire,* UK.

False teeth (individual porcelain/steel springs)
Claudius Ash, US.

Giant telescope
William Parsons, *3rd Earl of Rosse, astronomer, at Birr Castle, Parsonstown, Ireland. It had a* 72-*inch reflector and was* 54 *ft long, and he used it to discover the spiral nebulae.*

Horizontal turret lathe
Stephen Fitch, *Middlefield, Connecticut,* USA.

Saxophone, 1844

1845

Hydraulic crane
William George Armstrong, UK
inventor and solicitor.
Mustard mill (steam-powered)
Jeremiah *and* James Colman, *Stoke Mill, Norwich,* UK.
Number-stamping machine
William Shaw, UK.
Panoramic camera
F von Martens, *France. It was particularly useful for city scenes and school photographs.*
Rotary printing press
Richard M Hoe *of New York,* USA. *It was first used in 1847 on the Philadelphia Public Ledger.*
Rubber bands (vulcanised rubber)
Stephen Perry *of Messrs Perry &*

Co., rubber manufacturers, London.
Steam pile-driver
James Nasmyth, *Scottish engineer.*
Typewriter ribbon feed system
Alexander Bain, *London.*

1846
Adams patent omnibus.
Computing telegraph
J E Fuller *of New York,* USA, *combining Palmer's circular slide rule with his own time telegraph.*
Cup anemometer
Thomas R Robinson, *Irish astronomer.*
Ether vapour (anaesthetic)
Dr William Thomas Green Morton *of Massachusetts General Hospital,* USA. *Having tried it on his pet spaniel, then on his friends, then on himself, then on a patient at the hospital, Morton patented his discovery as 'letheon'.*
Gun cotton
Professor Christian Friedrich Schönbein *of Basel, German chemist, and* Rudolph Christian Böttger – *working independently of each other.*
Hexameter machine (automated poetry producer)
exhibited at the Egyptian Hall, Piccadilly, London. 86 wheels of all shapes and sizes, visible through a glass window, churned out 1440 hexameter verses per day. It was a complete hoax.
Neptune (planet)
Johann Gottfried Galle *and* Heinrich Ludwig d'Arrest *at the Berlin Observatory.*
Parchment paper (vegetable based)
Figuier *and* Poumarède *of Paris.*
Protoplasm (full investigation)
Hugo von Möhl, *Professor of Botany at Tübingen, Germany.*
Sponge rubber
Charles Hancock, *London.*
Vertical casting of iron pipes
D Y Stewart, *Scottish engineer.*

1847

Carbolic acid
Auguste Laurent, *France, by distilling pit coal.*
Chloroform anaesthetic
Jacob Bell *of London and* Dr James Young Simpson, *Scottish obstetrician, Head of the Midwifery Department of Edinburgh University. Although at first opposed on medical and theological grounds, Simpson's method was finally given official approval in 1853 after he had administered it to Queen Victoria during the birth of Prince Leopold.*
Double-decker bus
Adams & Co. *of Bow, London, for the Economic Conveyance Co. with accommodation for fourteen upper-deck passengers, sitting back-to-back on the roof.*
Ether inhaler
Attlee & Co, UK.
'Fish joint' (for uniting the ends of rails)
William Bridges Adams, *London engineer.*
Flute (present-day type)
Theobald Boehm, *distinguished Munich flautist, and his friend, the physicist* K F E Schafhäutl.
Iris (asteroid)
John Russell Hind, *astronomer, Royal Observatory, Greenwich, London. The first of eleven such asteroids he discovered.*
Kymograph
Karl Friedrich Ludwig, *physiologist, Marburg, Germany. It traced pulse beats on a graph.*
Nitro-glycerine
Ascanio Sobrero, *chemistry student, Italy.*
Railroad bridge truss
Squire Whipple, US *engineer.*
Rocket plane (experimental)
Ernst Werner von Siemens, *Berlin.*
Saccharimeter
Soleil, *Parisian optician, for determining the amount of sugar in solutions – using rotary polarised light.*

Steel gun
Alfred Krupp *at his cast steel factory at Essen, Rhenish Prussia.*
Taxi-meter cab
Patent Mile Index *fitted to a London cab.*
.22 rifle
Nicholas Flobert, *Parisian arms manufacturer.*
Vulcanised toy balloons
J G Ingram, *London.*

1848

Alpaca as umbrella-covering material
William Sangster, UK.
Chewing gum (spruce-based)
John Curtis *on a Franklin stove in the kitchen of his home at Bangor, Maine, USA. He called his product 'State of Maine Pure Spruce Gum'.*
Concrete boats
Joseph Louis Lambot, *French horticulturalist, used the method for rowing skiffs. He also made flowerpots in the same way – plastering sand-cement mortar over a framework of iron bars and mesh.*
Corliss valve (steam engine)
George H Corliss, *self-taught engineer from Rhode Island, New York, USA.*
Department store
Alexander Turney Stewart, *Irish ex-schoolmaster on lower Broadway, New York, USA. Called the 'Marble Dry Goods Palace'.*
Folding propeller (marine)
John Seaward, *civil engineer, Millwall, UK.*
Milling machine
F W Howe *for the Robbins & Laurence Co. of Vermont, USA.*
Telekouphonon
Francis Whishaw, *civil engineer of Hampstead, London. A speaking telegraph consisting of piping of gutta-percha, caoutchouc, glass or earthenware with a terminal mouthpiece of ivory, bone, wood or metal, it was used for long-distance*

communication in dockyards and large establishments.

1849

Amines
Karl Adolf Würtz, *Professor of Organic Chemistry, Mineral Chemistry and Toxicology at the Paris School of Medicine.*

Artificial leather-cloth
Messrs J R *and* C P Crockett *of Newark, USA. Unbleached cotton was coated with a mixture of boiled linseed oil and turpentine, then coloured.*

Benzole (or benzene)
Charles B Mansfield, *student of A W von Hofman at the London College of Chemistry. Distilling it from coal tar, Mansfield was the first to produce benzene on a large scale. Tragically he died six years later from burns received during a further experiment on it.*

Bloomers (women's trousers)
Mrs Amelia Bloomer, *(née Jenks) of New York, USA.*

Bowler hat
Thomas *and* William Bowler, *felt-hat makers of Southwark Bridge Road, London, as commissioned by Lock & Co. of St James's, Piccadilly for their customer, William Coke of Holkham, Norfolk.*

Building (cast-iron construction)
James Bogardus *for his foundry, New York City, USA.*

Fire annihilator
T Phillips, *to put out a fire using steam and carbonic acid.*

Methyl gas
Simultaneously by Edward Frankland, *organic chemist, Royal College of Engineers, UK, and* Hermann Kolbe, *Germany.*

Milk (condensed or evaporated)
Gail Borden, *near New York, USA. A factory went into operation in 1851 and successful manufacture was established by 1856. Cecil Borden invented the meat biscuit in 1850.*

Polar clock
Professor Charles Wheatstone – *an optical apparatus whereby the hour of the day was found by the polarisation of light.*

Safety pin
Walter Hunt, US, *and* Charles Rowley, *Birmingham,* UK.

Shrapnel shell (with time-fuse and diaphragm)
Captain Edward Boxer, *British Royal Artillery, at the Royal Laboratory, Woolwich,* UK.

Telephone (experimental)
Antonio Meucci, *Italian engineer working in Havana, Cuba.*

Velocity of light
Armand-Hippolyte-Louis Fizeau, *French physicist.*

1850

Binaural stethoscope
G P Caniman, *New York,* USA.

Derricks
A D Bishop, *of New York,* USA.

Foucault's Pendulum
Jean Bernard Léon Foucault, *French physicist, to prove that the earth rotates.*

Grain binder
John E Heath, US. *The first in which twine was used.*

Jeans (trousers)
Levi Strauss, *Bavarian immigrant to San Francisco, as hard-wearing work trousers for Gold Rush miners.*

Machine-made watch (interchangeable parts)
Aaron Dennison *and* Edward Howard *of Roxbury, Massachusetts,* USA. *It was later called the 'Waltham'.*

Mustard powder (in tins)
J & J Colman, *Stoke Mill, Norwich,* UK.

Myographion
Hermann von Helmholtz, *Professor of Physiology at Königsberg, Germany: a machine for determining the velocity of the nervous current.*

Photographic slides
Frederick Langenheim,

Philadelphia, USA.
Refrigerator
simultaneously by James Harrison,
newspaper publisher, at Rodney Point,
Victoria, Australia, and Alexander
Catlin Twining *at Cleveland, Ohio,*
USA.
Stearine candles
Messrs Price, *Battersea, London.*
This was achieved by the steam-
distilled saponification of fats.
Submarine lamp
Siebe & Gorman, *London.*
Teledynamic transmitter
M Hirn. *He assembled water-wheels,*
endless wires and pulleys for using the
power of waterfalls at a distance.
Thirteen-month calendar
Auguste Comte, *French philosopher*
and socialist, in his 'System of Positive
Politics'. Each month was dedicated
successively to Moses, Homer,
Aristotle, Archimedes, Caesar, Paul,
Charlemagne, Dante, Gutenberg,
Shakespeare, Descartes, Frederic and
Bichat; an eminent person was
commemorated every day.
Tubular suspension bridge
Robert Stephenson *and* William
Fairbairn, *engineers – 1 mile south of*
the Menai Suspension Bridge, and
called 'The Britannia'.
Vortex turbine
James Thomson, *Scottish engineer.*

1851
Carpet loom
Erastus Brigham Bigelow, *inventor,*
after six years' work at Lowell,
Massachusetts, USA.
Carriage odometer
William Grayson, *for measuring the*
distance covered by carriages.
Collodion process
Frederick Scott Archer, *portrait*
sculptor, UK, had learnt calotype
photography in order to have portraits
of his sitters as studies. His improved
Ambrotype soon ousted the calotype.
Colour-printing machine
G C Leighton, *using aquatinted plates*

and electro-typed silver and copper
surfaces to obtain purity of colour.
Compound ureas
Karl Adolf Würtz, *chemist, École de*
Médecine, Paris.
Cross-channel telegraph cable
Professor Charles Wheatstone,
taking up Faraday's suggestion to
insulate with gutta-percha, and
assisted by Joseph Brett. *It was laid by*
the steamship 'Goliath'.
Dairy ice-cream (mass-
produced)
Jacob Fussell, *dairyman of*
Baltimore, Massachusetts, USA.
Doppler's Principle
Christian Doppler, *Professor of*
Physics at Vienna, Austria.
High-speed flash photography
Henry Fox Talbot *of Laycock Abbey,*
UK. *1/100,000-second exposure,*
electric spark photo of a page of 'The
Times', attached to a spinning
cylinder.
Ophthalmoscope
Herman von Helmholtz, *physician*
and physicist, Germany. Initially the
machine directed a beam of light into
the eye by means of a mirror in which
there was a tiny lens aperture through
which an observer could study the
interior of the human eye.

Helmhotz's opthalmoscope

Printing machine
Thomas Nelson, *Edinburgh
publisher, whereby paper was printed
using curved printing plates affixed to
revolving cylinders.*
Quadricycle (man-pedalled)
Willard Sawyer *of Deal, Kent,* UK.
**Stereoscopic still camera
(lenticular)**
Louis Jules Duboscq, *France, as
proposed by* Sir David Brewster, UK.
*Binocular version was introduced two
years later by* J B Dancer *of
Manchester,* UK *and* A Quinet *of
Paris.*
Vulcanite/ebonite (hard rubber)
Nelson Goodyear, US, *and* Charles
Macintosh, UK.
Vulcanite dentures
in France.
**Wrought iron manufacture
(fuel-less pneumatic process)**
William Kelly, *ironmaster, Kentucky,*
USA.

1852
Airship (steam-powered)
Henri Giffard, *aeronautical engineer,
Paris. The 144-ft dirigible, powered by
a 3 hp steam engine, voyaged 17 miles
at 6 mph.*
Compressed air caisson
Isambard Kingdom Brunel, *to assist
construction of a bridge at Chepstow,
Monmouthshire,* UK. *Fourteen cast-
iron cylinders filled with concrete or
brickwork, used to lay foundations
under water.*
Cork lifebelt (multi-section)
Captain John Ross Ward RN *of the
National Life Boat Institution. 800
units were supplied to life boat crews
and remained standard issue from
1856 to 1904.*
Eau Grison (fungicide)
M Grison, *French gardener.*
External screw top (for bottles)
François Joseph Beltzung, *engineer,
Paris, France.*
Fluorescence
Professor George Gabriel Stokes,
mathematician and physicist, Fellow

of Pembroke College, Cambridge.
Glider (man-carrying)
Sir George Cayley *of Brompton Hall,
near Scarborough, Yorkshire,* UK. *The
fixed-wing glider, piloted by Sir
George's coachman, became airborne
after being towed by manpower down a
hill against a slight breeze. After
touchdown, the terrified coachman is
reported to have given in his notice,
shouting 'Please sir – I was hired to
drive, not to fly!'*
Goods elevator
Elisha Graves Otis, *master mechanic
of a New York bedstead company.*
Lutetia (asteroid)
Hermann Goldschmidt, *German
astronomer in Paris. This was the first
of fourteen asteroids he discovered by
poking small telescopes through his
attic window.*
Microfilm
John Benjamin Dancer *of
Manchester,* UK, *using collodion film.*
'Odic force'
Karl Baron von Reichenbach, *who
discovered a new force intermediate
between electricity, magnetism, heat
and light and recognisable only by the
nerves of sensitive persons.*
Rake (mechanical)
Samuel Johnston, US.
Rotary hook (sewing device)
Allen B Wilson, US, *financed by
Nathaniel Wheeler.*
Ship (caloric-engined)
Jon Ericsson, *Swedish inventor
working in the* USA.
**Thesaurus of English words and
phrases**
Peter Mark Roget, *scholar and
physician, London.*
Thetis (asteroid)
R Luther, *astronomer; the first of
twenty such asteroids he discovered.*
Water ejector pump
James Thomson, *Scottish engineer.*

1853
Chuck lathe
E K Root *at the Colt Armory,* USA.

Cigarette facory (steam-powered)
Don Luis Susini *in Havana, Cuba.*
Capacity rose to 2,583,000 per month.

Cutting-out machines
Frederick Osbourn, *and manufactured by Messrs Hyams for cutting out clothes.*

Dark-field microscope condenser
Francis H Wenham, UK.

Duplex telegraphy
Dr Wilhelm Julius Gintl, *Austria.*

Gas engine (internal combustion)
Eugenio Barsanti *and* Felice Matteucci *of Florence, Italy. The first working unit was installed at the Maria Antonia Railway Station, Florence, three years later.*

Gaslit cystoscope (medical).

Hypodermic syringe (silver construction)
Charles Gabriel Pravaz, *French surgeon of Lyons, France.*

Hypodermic syringe
Alexander Wood, *physician and lecturer, Edinburgh University,* UK, *to administer morphia.*

Multiplex telegraphy
Moses Gerrish Farmer, *of Salem, Massachusetts,* USA.

Panopticon of science and art
T H Lewis, *architect, in Leicester Square, London – for lectures and musical performances. It included a very large electrical machine and battery. The speculation failed and the business was sold off in 1857.*

Railway carriage (corridor approach)
Messrs Eaton & Gilbert *of Troy, New York for the Hudson River Railroad.*

Rocking chair (revolving with adjustable seat angle)
Peter Ten Eyck, *New York,* USA.

Saltaire (model village)
Sir Titus Salt, UK *alpaca wool manufacturer and benefactor. He built the village for his employees, around*

his factories beside the Aire River.

Steam-powered fire engine (self-propelled)
A B Latta *of Cincinnati,* USA.

Trotman's anchor
John Trotman *of Dursley, Gloucestershire,* UK.

1854

Air filters and respirators (charcoal-based)
Dr John Stenhouse, *chemical lecturer, St Bartholomew's Hospital, London. First used at the Mansion House, London.*

Aluminium (chemically produced)
Henri Sainte Claire Deville, *France, based on the pioneer work of Davy, Oersted and Wöhler. The first aluminium factory went into operation at Glacière in 1856.*

Annular-sail wind pump
Daniel Halliday.

Boolean algebra
George Boole, *Professor of Mathematics, Queen's College, Cork, Ireland.*

Cartes de visite (photographic)
Disdéri *of Paris, who succeeded in taking eight photos on one negative. They were popularised in 1857 when the Duke of Parma had his portrait, as taken by Ferrier in Nice, France, placed on his visiting cards.*

Chemical wood pulp
Watt *and* Burgess, *who used caustic soda to digest wood pulp directly under pressure to make paper.*

Compound marine engine
John Elder & Co., *marine engineers, Clyde,* UK – *for* SS *'Brandon'.*

Cornflour
John Polson, *later of Brown & Polson Ltd,* UK.

Electric loom
Cavaliere Gaëtano Bonelli *of Turin – employing magnets and electro-magnets in weaving, attempting to supersede the costly Jacquard system of cards.*

1854

Electro-magnetic manufacture
Breguet, *France – of mathematical instruments.*

Four-motion feed (sewing device)
Allen B Wilson, US.

Giant train shed
New Street Station, *Birmingham*, UK. *The span of its iron roof was 211 ft.*

Letter-printing telegraph
David Edward Hughes, *London-born Kentucky music professor.*

Palaeozoic fossils
Reverend Adam Sedgwick, UK *geologist.*

Paraffin lamp
John H *and* George W Austen, *New York*, USA.

Perforated postage stamps
UK.

Planimeter (constant length tracing arm)
Amsler. *12,000 such planimeters had been manufactured by 1884.*

Psychic pressometer
Dr Robert Hare, *Professor of Chemistry at Pennsylvania University*, USA, *to expose fraudulent 'table-tapping' mediums. Several mediums 'forced' the machine to register the equivalent of 18–40 lb, ultimately collapsing the apparatus. When a convinced Hare attempted to present his evidence to the American Association for the Advancement of Science, he was ridiculed as senile.*

Telephone (experimental)
Charles Bourseul, *French telegraphy engineer.*

1855

Armoured vehicle (steam-powered)
Cowan, UK.

Bichromate battery
Heinrich Ruhmkorff, *German physicist.*

Bunsen burner
Robert Wilhelm Eberhard von Bunsen, *Professor of Chemistry at Heidelberg. In that year Heidelberg* came to be lit by gas. Bunsen *researched a controllable-flame burner to use in his laboratory, the former refectory of an ancient monastery.*

Cased-in water turbine
James Bicheno Francis, UK *émigré engineer to the* USA.

Chronograph (stopwatch)
Edward Daniel Johnson, *watch manufacturer of London.*

Cupola (turret) ship
Captain Cowper Coles, *British naval officer. Several such ships were constructed by E J Reed, including 'Sovereign', 'Monarch' (armour-clad) and 'Captain', which sank in a storm in the Bay of Biscay with the loss of 472 lives, including the inventor.*

Dry-cleaning (clothes)
Jean Baptiste Jolly, *French dyeworks owner in Paris, France. His maid accidentally upset a paraffin lamp on a tablecloth and Jolly noted that the area soaked became cleaner. Operating from his dyeworks, he offered the new process as 'dry-cleaning' as opposed to the soap and water process.*

Fire-fighting vessel (steam-powered)
James Braidwood, *Superintendent of the London Fire Engine Establishment, based on an idea by* John Braithwaite, *engineer*, UK.

Glider (sea-going)
Jean Marie Le Bris, *French sea captain, who made a glide of 200 metres near Douarnenez in a boat-shaped machine, patterned after the albatross.*

Locomotive speedometer
A W Forde.

Pedal-operated machine kettledrum
M Gautrot, *Parisian instrument-maker.*

Portable prospector (hand, horse or steam power)
Thomas Oultram *and* Jacob Braché, *engineers, Melbourne, Victoria, Australia. Australian Patent Number 2.*

Powdered milk (industrial manufacture)
Grünwald, Germany.
Safety match
Johan Edvard Lundström *of Jönköping, Sweden, using red phosphorous for the striking surface as suggested ten years before by* Gustave Pasch *of Sweden and* Professor R Bottger *of Frankfurt-on-Main, Germany.*
'Scaphandre' (diving apparatus)
Joseph Martin Cabirol *of Narbonne, France.*
Smith & Wesson revolver
Daniel Baird Wesson *at Springfield, Massachusetts, USA.*
Spinal anaesthesia
J L Corning, US, *who used a solution of cocaine to inject into nerve endings.*
Timber-bending machine
T Blanchard *of Boston, Massachusetts.*
'Very Powerful Electric Machine' (Leyden-jar-powered)
Dr Henry Minchin Noad *at the Panopticon, Leicester Square, London.*
Z-crank steam engine
Morton & Hunt, *Glasgow, UK.*

1856
Aerated bread
Dr John Dauglish, *Edinburgh. Marketed by the* ABC *(Aerated Bread Co.) at their London tea-rooms in 1861.*
Aniline dye
William Henry Perkin, *aged eighteen, while studying at the London College of Chemistry under the German chemist, A W von Hofman. The following year, with capital from his father and brother, Perkin set up a chemical works to turn out 'aniline purple', later called 'mauve'.*
Bessemer converter
Henry Bessemer, UK *engineer, who set up his own steelworks at Sheffield to promote the process.*

Clockwork toy train
George Brown, US *clockmaker.*
Equatorial sextant
William Austin Burt, *surveyor, Mount Vernon, Michigan, USA.*
Fuchsine dye
Wladyslaw Natanson, *Poland.*
Glycogen
Claude Bernard, *French physiologist.*
Glycol
Charles-Adolphe Würtz, *Dean of the École de Médecine, Paris.*
Parkesine
Alexander Parkes, *chemist of Birmingham, UK – the prototype celluloid. The Parkesine Co. Ltd was formed in 1866, but lasted only two years.*
Patent safety railway points and signals (interlocking)
John Saxby, *railway engineer of Brighton, Sussex, UK. In 1862 he went into partnership with John Stinson Farmer to manufacture from a factory in Kilburn, London.*
Practical gas fire
Pettit *and* Smith, UK.
Sarrusphone
Sarrus, *French military-bandmaster, together with* P L Gautrot, *Parisian instrument-maker – 'a brass oboe'.*

1857
Compressed-air drilling (tunnels)
for the Mount Cenis Tunnel, Switzerland.
Concrete mixer
for construction of a bridge over the River Tisza at Szeged, Hungary.
Domestic sewing machine (production model)
Isaac Merritt Singer *of Pittsdown, New York, USA. It was based on his treadle-operated lock-stitch machine of 1851.*
Lamplit street (electric power)
Messrs Lacassagne et Thiers *in the rue Impériale, Lyons, France.*
Laryngoscope
Johann Nepomuk Czermak, *Polish*

physiologist working in Vienna.
Passenger elevator
Otis Steam Elevator Co. *in E V Haughwout & Co., a five-storey china shop in Broadway, New York,* USA. *It raised six people at 40 ft per minute.*
Phonautograph
Léon Scott de Martinville, *French painter – to analyse sound waves.*
Photoheliograph
simultaneously by Sir John Herschel *at Kew Laboratory, London, and* H van der Weyde, US *artist.*
Silicon nitride
independently by Henri Etienne Sainte-Claire Deville, *French chemist, and* Friedrich Wöhler, *German chemist.*
Steam-powered franking machine.
Vibration microscope
Jules Antoine Lissajous, *Professor of Physics at the College St Louis, Paris. This showed visually the 'Lissajous figures'.*
Wood pulp (alkaline process)
Houghton *at Cone Mills, Lydney, Gloucester,* UK.

1858
Burglar alarm
Edwin T Holmes, US. *Installed in Boston, Massachusetts.*
'Cell-from-cell' theory
Rudolf Virchow, *Professor of Pathological Anatomy at Berlin.*
Donati's Comet
Dr Giovanni Battista Donati *of Florence, Italy.*
Gas-lit railway carriages
Eastern Railway Co., *France.*
Harvester
Charles Wesley Marsh, *and his brother* William, *Illinois,* USA.
Lighthouse (magneto-electric)
Professor Holmes. *South Foreland lighthouse, Dover,* UK.
Mechanised washtub and dolly
Hamilton Smith, *Pittsburg,* USA.
Möbius strip
Professor August Möbius, *whilst*

investigating polyhedra. Not published until 1865.
Pencil with attached eraser
Hyman Lipman *of Philadelphia,* USA.
Pistolgraph (still camera)
Thomas Skaife. *Its 'snap-shooting' inventor was nearly arrested for aiming it at Queen Victoria.*
Quadrivalent nature of carbon
Archibald C Couper, *Scottish chemist.*
Seismometer
Robert Mallet, *civil engineer, Dublin, Ireland.*
Shoe-sole-sewing machine
Lyman R Blake, *Abington, Massachusetts,* USA. *It was perfected soon after with financial assistance from Gordon McKay.*
Steam plough
John Fowler, *agricultural engineer, and* Jeremiah Head, UK.
True atomic and molecular weights
Stanislao Cannizzaro, *Professor of Chemistry in Genoa, Italy.*

1859
Accumulator/storage battery
Gaston Planté, *France, using lead-acid. Based on* Johann Wilhelm Ritter's *discovery of the secondary cell, as early as 1803.*
Aerial screw propeller
J B Lassie, *whose model for an aerial ship was submitted to the Paris Academy of Sciences, although the full-scale craft was never built.*
Aniline yellow dye
Peter Greiss, *a German working in a brewery in Burton-on-Trent,* UK.
Binoculars (central focus dial)
A A Boulanger, *France.*
Breech-loading gun (converted)
Jacob Snider, *wine merchant, Philadelphia,* USA. *His invention was finally adopted by the government in 1866 and an order placed for 100,000 breechloaders.*
Cathode rays
Julius Plücker, *mathematical*

Steam organ

physicist at Bonn, Germany.
Horizontal-arch reservoir dam
Zola, *French engineer, at Aix-en-
Provence, based on recommendations
by* F Delocre.
**Merchant ship (bessemer steel
construction)**
Miniature camera
T Morris *of Birmingham,* UK.
Oil well
'Colonel' Edwin L Drake *at
Titusville, Pennsylvania,* USA, *whilst
working for George Bissel, whose small
company had been marketing lamp oil
made from crude oil obtained from
surface seepages. Success at Titusville
was achieved when the hole partly
filled with oil at a depth of* $69\frac{1}{2}$ ft.
'On the origin of the species'
Charles Robert Darwin, UK
*naturalist, following 25 years'
research.*
Pneumatic dispatch
*for conveying letters and parcels
through tubes by means of atmospheric
pressure and a vacuum.*

Primitive fluorescent lamp
Edmond Becquerel, *France, who
placed fluorescent materials inside a
Geissler discharge tube.*
Pullman cars (railway)
George Mortimer Pullman, *Chicago,
Illinois,* USA. *He made his first sleeping
cars using two old day coaches divided
into ten compartments. In 1864 he
patented the folding upper berth. In
1867 the Pullman Palace Car
Company was formed; in 1868 dining
cars were introduced.*
Spectroscope
Gustav Robert Kirchoff *and* Robert
Wilhelm Bunsen, *German physicists
working in Heidelberg.*
Steam organ (calliope)
Arthur Dennis, US. *Powered by a 30
hp engine, developing a steam pressure
of 125 lb per square inch, kalliopes
were installed at St Louis and New
Orleans.*
Steam road-roller
Louis Lemoine, *France.*

Wide-angle lens (camera)

Thomas Sutton, UK editor of

'Photographic Notes'.

Caesium

Robert Wilhelm Bunsen and Gustav Robert Kirchoff, using spectrum analysis.

Debusscope (pattern-making device for calico-printers)

France.

Linoleum

Frederick Walton, who built his first factory at Staines, Middlesex, UK. It was patented by a rubber manufacturer to replace Kamptulicon, an oil-rubber product.

Press-stud fastener

John Newnham, UK.

Solar boiler (for distillation)

August Mouchot, France.

Synthetic ruby

Edmond Frémy, French chemist.

Winchester repeating rifle

B Tyler Henry, chief designer for Oliver Fisher Winchester, US gun and ammunition manufacturer.

Enamel-mosaics

Dr Antonio Salviati, while reviving the glass industry of Murano, Venice.

Esparto pulp

Thomas Routledge of Eynsham Paper Mills, Oxford, UK, from esparto grass grown in the Mediterranean lands. The innovation proved so successful that in 1862, 1000 tons of esparto grass were imported into the UK.

Harpoon log

Thomas Walker, UK.

Hydrofoil boat

Thomas Moy, UK – on the Surrey Canal between Rotherhithe and Camberwell.

Kinematoscope (stereoscopic photos)

Coleman Sellers of Philadelphia, USA.

Mechanical road transport (army use)

R E Crompton, whilst at Harrow School, UK. His 'Blue Belle' was shipped out to India in 1867, where Crompton was serving with the Rifle Brigade.

Motor speech centre of the brain

Pierre Paul Broca, surgeon to the Paris hospitals.

Photography (underwater)

William Thompson of Weymouth, UK. His initial experiment was made 5 years previously, by lowering a watertight box from a boat in Weymouth Bay.

Pneumatic drill

Germain Sommelier, chief engineer on the Mont Cenis railway tunnel project.

Postcard

John P Charlton, Philadelphia, USA.

Rubidium

Robert Wilhelm Bunsen and Gustav Robert Kirchoff, Germany, using spectrum analysis.

Telephone (partially articulate)

Philipp Reis, physics teacher at a private school near Frankfurt-am-Main, Germany, for the enlightenment of his pupils. Although Reis lectured on and demonstrated his machine publicly, he was unable to realise its full potential, dying aged only 40 after a long illness which eventually robbed him of his voice.

Thallium

William Crookes, chemistry lecturer and editor of 'Chemical News', London.

Transatlantic oil cargo.

Universal milling machine

Joseph R Brown, US, to manufacture musketry parts for the US Civil War.

Auriscope

John Brunton.

Bread-making machine

Ebenezer Stevens of Cheapside, London.

Ironclad ship
Jon Ericsson's *'Monitor'*, *complete with a gun turret, as invented by* T R Timby, US, *some twenty years before. That spring, 'Monitor' engaged with the Confederate ironclad 'Merrimac' at close quarters. Each hit the other with shells over twenty times, and they withdrew in stalemate.*
Machine-gun
Richard Jordan Gatling *of North Carolina*, USA. *His manually rotated ten barrels fired bullets, gravity-fed from a drum at the rate of 350 rounds per minute.*

Cartridges for Gatling's machine gun

Milling machine
Joseph Brown *of Brown & Sharpe Co., Providence, Rhode Island*, USA.
Phonoscope
Dr Rudolph Koenig *of Paris, to test the quality of musical strings.*
'Reciprocal electrophorus'
Cromwell Fleetwood Varley, *electrical engineer*, UK.
Rubber stamps
John Leighton FSA, *but patented two years later in the name of Alfred Leighton. They were made of vulcanised india-rubber.*

Steel-hulled merchant ship (transatlantic)
paddle-steamer, 'Banshee', 325 tons, built in Liverpool as a blockade-runner for the US *Civil War.*
Washing machine
100 garments were washed in a few minutes by steam and dried by a centrifugal machine at a New York hotel.

1863
Anti-cyclone (meteorology)
Francis Galton, *scientist and explorer,* UK.
Dental drill (clockwork-powered)
George Fellows Harrington, *subsequently manufactured at his own factory in Lind Street, Ryde, Isle of Wight*, UK.
Harpoon gun (practical)
Sven Foyn *of Tönsberg, Norway. It was fitted aboard the whaler 'Spes et Fides' in 1864.*
Indium
Ferdinand Reich, *German physicist,* and Hieronymous Theodor Richter, *German chemist.*
Medical photography
Dr Henry Goode Wright, *London.*
Motor carriage (coal gas)
Etienne Lenoir, *France, who fitted one of his 1½ hp gas engines to a carriage and used to take 1½ hours to drive 6 miles from his home to his works.*
Multi-plate clutch (locomotive)
Thomas Aldridge Weston, *mechanical engineer, Birmingham,* UK.
Pepper's ghost
John Henry Pepper, *analytical chemist at the Royal Polytechnic Institution of London. Pepper adapted his invention from a design by* Henry Dircks, UK *civil engineer, who used mirrors and lenses to 'project' a ghost.*
Phonograph
Fenby. *A machine to be attached to pianos, etc. by which any music played was written down on blank paper, since*

1863

it ruled and printed the notes simultaneously.

Regenerative smelting oven (Siemens-Martin process)
initiated in 1856 by William *and* Friedrich Siemens, *brothers, for glass and steel manufacture, and perfected by* Emile *and* Pierre Martin, *father and son, in Sireuil, France.*

Roller-skates (four-wheeled)
J L Plimpton, *New York,* USA.

Sodium carbonate production process
Ernest Solvay, *Belgian chemist.*

Sphygmograph
Jules-Etienne Marey *of Paris, for investigating disease by showing the state of the pulse.*

Submarine (explosively charged)
Horace L Hunley's *'David', propelled by an eight-man, hand-cranking crew. It was destroyed while blowing up a Federal vessel in the* US *Civil War.*

TNT (explosive)
J Wilbrand, *Germany.*

Underground steam railway
a broad gauge line from Victoria Street to the Great Western Railway at Paddington, London. The carriages and stations were gas-lit. The exhaust from the steam engines was returned to the tanks by big pipes from the cylinders, based on the idea of George Pearson, *solicitor.*

1864

Anthropoglossus
an alleged automaton talking machine was exhibited at St James's Hall, London, but proved to be a complete hoax.

Driving chains
James Slater, *owner of a small textile-machine factory at Salford,* UK. *In 1880,* Hans Renold, *a Swiss who had acquired Slater's factory, patented the bush-roller chain to give a much greater load-bearing surface.*

Electromagnetic wave

transmission
Dr Mahlon Loomis, *who succeeded in transmitting and receiving electromagnetic waves without wires over 14 miles at Loudon County, Virginia,* USA.

Narrowboat (steam-powered)
Grand Union Canal Company's 'Dart', UK.

Nitro-glycerine explosive
Alfred Nobel, *Swedish inventor and manufacturer in St Petersburg.*

Open hearth furnace (pig and scrap melting)
Emile *and* Pierre Martin, *father and son, in Sireuil, France.*

Pasteurisation
Louis Pasteur, *Dean of the Faculty of Sciences, Lille, France, the centre of the brewing industry.*

Louis Pasteur

Photography (using magnesium flash)
Alfred Brothers *of Manchester,* UK.

Smokeless powder
J F E Schultz, *Germany – very effective for hand grenades and shotguns.*

Solar-powered steam engine
August Mouchot, *French engineer.*
This was the prototype for a series he
built during the next fifteen years.

1865
Aérophore (compressed-air
diving apparatus)
Benoît Rouquayrol, *mining engineer,*
and Auguste Denayrouze, *French*
naval officer.
Automatic turret lathe
Christopher M Spencer, US.
Canoe (European pleasure)
John MacGregor, *who persuaded*
Searle's of Lambeth, London, to build
him his first Rob Roy canoe, influenced
by his seeing the canoes of North
America and the Kamschatka.
Cyclic structure of benzene
August Friedrich Kekulé, *German*
Professor of Chemistry at Ghent.
'Duplex burner' (oil lamp)
James *and* Joseph Hinks,
Birmingham, UK.
Electric arc welding
Wilde.
Ferryboat (air-cushioned)
running between New York and New
Jersey, USA. While the vessels went
with less power applied to the paddles,
the power required to provide
compressed air along channels in the
keel was just about equal to what was
saved in driving the paddles – so the
service was discontinued.
Genetics
Gregor Johann Mendel, *Austrian*
biologist and Augustinian Abbot of
Brun, who combined his love of
mathematics with his knowledge of
plants.
Motorboat (gas-engined)
Jean Joseph Etienne Lenoir, *France.*
Oil pipeline (commercial)
Samuel van Syckel, *pioneer refiner –*
about 6 miles, from a flush oil field at
Pithole to a railroad loading point at
Miller Farm, Pennsylvania, USA.
Pedal bicycle (front-wheel-drive

production model)
Pierre Michaux, *Parisian coach and*
perambulator manufacturer, and his
son, Ernest. *Over 400 velocipedes sold*
that year.
Reinforced concrete
W B Wilkinson *of Newcastle-upon-*
Tyne, UK *who built a house in which*
the concrete beams were reinforced
with wire colliery rope.
Spinnaker yachting sail
William Gordon, *sailmaker of*
Southampton, UK, *who first used a*
prototype in his yacht 'Niobe' in a
match race off Cowes, Isle of Wight.
Steam road-roller
Thomas Aveling, *former farmer and*
agricultural mechanic from Kent, UK.
For two years Aveling's machine was
attacked and banned. In 1867 the City
of Liverpool ordered the first steam-
roller.
Tinplated food cans.
Uninflammable gunpowder
Mr Gale, *a blind gentleman from*
Plymouth, Devon, UK, *by combining*
gunpowder with finely powdered glass
which could be readily separated by a
sieve when the powder was to be used.
Gale's Protected Gunpowder
Company was formed in 1865 and
wound up in 1867.
Web-offset printing machine
William Bullock, *Pittsburg,*
Pennsylvania, USA, *who lost his life*
soon after in an accident with one of his
presses.
Western ('10 gallon') hat
John B Stetson, US.
Yale cylinder lock
Linus Yale Jr, US – *enabling the mass-*
production of locks to become a reality.
Sadly Yale died, aged only 47, from
heart failure in 1868, just before his
Lock Manufactory went into
operation (see overleaf).

1866
Aerial cablecar
Ritter, *so that people could inspect*
turbine installations on the Rhine,

1866

Yale Pattern night lock, *1865*

Germany. *It had a track of 101 metres, four steel-carrying cables and conveyed two people in one cabin. Its hauling rope was hand-cranked.*

Atlantic telegraph cable
Cyrus West Field, US, *with* Samuel Canning *and* Daniel Gooch, UK. *Over 3000 miles of cable were laid by the 'Great Eastern' from Valentia, Ireland to Heart's Content, Newfoundland. It was said that the engineer of the cable passed signals through it by means of a battery formed in a lady's thimble.*

Chassepot rifle
Alphonse Chassepot, *France. By 1867, 10,000 chassepots had been issued to French troops.*

Cigar-shaped steamship
William Louis *and* Thomas Winans, *formerly of Baltimore,* USA, *but resident in London; built by Hepworth and launched on the Thames,* UK.

Depressor nerve
Karl Friedrich Wilhelm Ludwig, *Professor of Physiology, Leipzig, with* Élie de Cyon.

Electro-magnetic road vehicle
E Poitevin.

Fire-extinguisher (carbon dioxide gas)
François Carlier, *French doctor – by mixing sodium bicarbonate with sulphuric acid.*

Ironclad gunboat (waterjet-propelled)
HMS *'Waterwitch'*, *British Admiralty – propelled by the hydraulic reaction arrangement introduced by J Ruthven.*

Key-opening canned food
J Ousterhoudt, *New York,* USA.

Self-acting dynamo
Werner von Siemens, *Germany.*

Torpedo (self-propelling)
Robert Whitehead, UK *engineer with a works at Fiume, near Trieste, Italy, in conjunction with Captain Giovanni Luppis of the Austrian Navy. Originally carried 19 lb of guncotton, measured 11 ft long and travelled at 7 knots.*

Visible speech (lip-reading)
Alexander Melville Bell, *Scottish-American educationalist, working in London.*

Wood pulp (prototype sulphite process)
B C Tilghmann *and* R Tilghmann, *brothers at the mills of W W Harding & Sons, Manayunk, near Philadelphia,* USA. *Unable to overcome engineering problems associated with their process, the Tilghmanns never promoted the idea commercially.*

1867

Amphibious velocipede (lever-operated)
William George Crossley, *dentist of Cambridge,* UK.

Automatic electric block railway
Thomas S Hall *on the New York and Harlem Railroad,* USA.

Automatic printing telegraph
Professor Charles Wheatstone, UK.

Barbed wire
Lucien B Smith *of Kent, Ohio,* USA.

'Bismarck' brown' (dye)
Carl Friedrich Philipp von Martius, *German naturalist.*

Double-decker railcars
Bidel, *France, for the Paris ring-line.*

Dynamite
Alfred Nobel, *inventor and manufacturer at Ammeberg, Sweden.*

Formaldehyde
August Wilhelm von Hofmann, *London-based German chemist.*

Mechanical 'writing ball'
Pastor Rasmus Hans Johan Malling Hansen, *Head of the Royal Deaf and Dumb Institute, Copenhagen, Denmark.*

Ring armature (generator)
Zénobe T Gramme, *Belgian scientist working in Paris.*
Time-switch
Dr Walter Christopher Thurger *of Norwich,* UK *– an eight-day clockwork 'gas controller'.*
'White coal' (hydro-electric power)
Aristide Bergès, *paper-maker, for his factory at Lancey, Isère, France.*

1868
'Air-hardening' tool steel
Robert Mushnet, UK.
Asbestos (industrial use)
200 tons of raw material were worked from an Italian mine, and it was discovered in Quebec, Canada. This heat-resistant, fibrous mineral was known and named by the ancient Greeks.
Boneshaker bicycle (steam-powered)
Sylvester Howard Roper *of Roxbury, Massachusetts,* USA, *and independently by* Pierre Michaux *and* L G Perreaux *of Paris.*
Compressed-air brake (railways)
George Westinghouse, US

Bicycle tuition

Cryptograph
Sir Charles Wheatstone, UK *– an apparatus for writing in cipher.*
Elevated street railway
Charles T Harvey, *civil engineer, New York,* USA. *It was called the 'El'.*
Gas geyser (mechanical water heater)
Benjamin Waddy Maughan, *London decorator (see overleaf).*
Granny Smith apple
Miss Smith, *Australia.*
Helium (in sun's chromosphere)
Edward Frankland, *Professor of Chemistry, Royal Institute, London,* with Joseph Norman Lockyer, *astronomer.*
Leclanché battery
Georges Leclanché, *France, as a dry-cell non-rechargeable primary battery.*
Margarine
Hippolyte Megé-Mouriés, *at the Ferme Impériale de la Faisanderie, Vincennes, France. His was the only entry in prize competition organised by Napoleon III for 'a suitable substance to replace butter for the Navy and the less prosperous classes'.*
Mechanical sheep-shearing machine (prototype)
James Andrew Higham, *compositor, Melbourne, Australia.*
Naval searchlight
Commander Philip Howard Colomb. *It was installed on 'HMS Minotaur', for use in 'Colomb's flashing signals'.*
Nephoscope
Karl Braun *– to measure the velocity of clouds.*
Plywood systems
John K Mayo, *New York,* USA. *Mayo's various patents show applications for bridges, canal locks, sewer's, tables and doors. He even suggested wet and dry heating to make the veneers more pliable before their compression.*
Pocket calculating machine
Charles Henry Webb, US, *'Webb's Adder'.*

1868

Maughan's gas geyser, 1868

Shellac mouldings
although recorded in very ancient writings, the first patents were taken out during this year.
Stapler
Charles Henry Gould *of Birmingham*, UK.
Steam man
New York – a figure constructed to drag a phaeton.
Street crossing signals
J P Knight, *railway signalling engineer, outside the Houses of Parliament, Westminster, London. The semaphore arms had red and green gas lamps for night-time. It was dismantled after it exploded, killing a policeman.*

1869
Aero-steam engine
George Warsop, *Nottingham*

mechanic, UK. *Compressed air was united with steam, saving 47 per cent fuel. A tug-steamer built for China was fitted with an aero-steam engine in 1870.*
Celluloid
John Wesley Hyatt *and his brother,* Isaiah Smith Hyatt, US. *Initially called collodion and used for coating billiard balls etc., it was very inflammable. It was also used for denture plates and toys – 'caveat emptor'!*
Chain-driven velocipede
Guilmet, *France.*
Colour photography (subtractive method)
simultaneously by Charles Cros *and* Louis Ducos du Hauron, *both of whom based their system on the Young-Helmholtz colour separation and mixing theory by superimposing three colour positive pictures on one another.*
Eugenics
Francis Galton, *scientist,* UK.
Gun (compressed steel)
Sir Joseph Whitworth, *industrialist and mechanical engineer, Manchester,* UK. *Its spiral polygonal bore gave it a considerably increased range.*
Hypodermic syringe (glass construction)
Luer, *France.*
Lithofracteur (stone-breaker)
Professor Engels *of Cologne, Germany, and manufactured by* Krebs. *A type of dynamite.*
Monorail-trolley 'dis-assembly' line
US *meat-packing firms in Cincinnati and Chicago.*
Nickel-plating.
Ominimeter
Eckhold, *German engineer, to replace chain measuring. It comprised a theodolite, a level, a telescope and a microscope.*
Pantograph
Francis Galton, *scientist for the Meteorological Office,* UK.
Periodic law
Dmitri Ivanovitch Mendeleyev,

Professor of Chemistry at St
Petersburg, Russia.
Rayon acetate
Paul Schützenberger, *physician and
Director of the Sorbonne Chemical
Laboratory, Paris.*
Rickshaw
Reverend Jonathan Scobie, US
*Baptist minister, to transport his
invalid wife about the streets of
Yokohama.*
Suez Canal
Ferdinand de Lesseps, *French
engineer, after a project lasting nine
years.*
Vacuum cleaner
Ives W McGaffey. *The manually
operated sweeping machine
incorporated a suction device for
surface cleaning.*
Xylonite
Daniel Spill, *former partner of
Alexander Parkes, inventor of
Parkesine,* UK.

1870
Bucket-wheel water turbine
Lester Allen Pelton, US *engineer,
California,* USA.
Chewing gum (chicle-based)
Thomas Adams, *Staten Island
photographer – whilst researching a
substitute for rubber. He began a
factory two years later.*
Circular ship
Admiral Popoff, *Russian Navy. It
had six propellers and two 11-inch
turret guns and was 101 ft in diameter.*
Fireman's respirator
Dr John Tyndall, *Irish physicist.*
Goods-wagon (steam-engined)
John Yule, *Glasgow,* UK, *for
transporting large marine boilers from
his works at Rutherglen Loan to the
Glasgow docks.*
Helicopter (steam-engined)
Horatio Phillips *of Battersea,
London. It failed to take off.*
**Pasigraphy (international
communication by numbers)**
Anton Bachmaier *of Munich. He*

*published a number-grammar of
German, French and English, enabling
4334 mental conceptions to be
communicated numerically.*
Phenophthalmoscope
Franciscus Cornelis Donders, *Dutch
oculist and Professor of Physiology at
Utrecht, for investigating eye-ball
movements.*
**Pneumatic underground
railway**
New York, USA.
Stock-ticker
Thomas Alva Edison *and* Franklin
Leonard Pope, *electrical engineers at
a workshop in Newark,* USA. *Some
months before, Edison had repaired the
ticker of the New York Gold
Reporting Company and been made
foreman by the company's president,
Dr Laws.*
**Washout water closet (all-
earthenware)**
Thomas W Twyford *of Hanley –
'The National'. 100,000 were sold and
in use within two decades.*
**Wheel (threaded spoke with
flange-headed nut)**
William Henry James Grout,
engineer, UK.

1871
Automatic Coupler (railways).
Bulb horn.
**Dry-plate photograph (gelatine
silver bromide emulsion)**
Dr Richard Leach Maddox, UK
physician and microscopist.
**Halftone process (photographic
reproduction)**
Carl Gustaf Wilhelm Carlemann,
Swedish engraver.
**Ordinary ('Penny-farthing')
bicycle**
James K Starley *of the Coventry
Machinist Company, with* William
Hillman. *Called the 'Ariel'.*
Paranormal isolation cage
William Crookes *and* Dr William
Huggins, *Fellows of the London Royal
Society, to test the paranormal powers*

1872

The 'Ariel' bicycle

of Daniel Home in laboratory
conditions. Crookes observed how an
accordion in a wire cage surrounded by
an electrical current could be made to
levitate and play tunefully without
Home physically touching it.
Phenolphthalein dye
Johann Friedrich Wilhelm Adolf von
Baeyer, *German organic chemist at the
Berlin Technical Institute.*
**Self-raising flour (chemical
raising agent).**
Toilet roll
Seth Wheeler, *New York,* USA.
Wind tunnel
Francis Herbert Wenham, UK *marine
engineer, interested in aeronautics. It
was built by Messrs Penn of London.*

1872
**Cooke's circular dividing
machine**
Frederick Cooke, UK.
Hydroplane (boat)
Reverend Charles Meade Ramus,
Rector of Playden, near Rye, Sussex,
UK – *for designing steamships of great*

*speed. He also advocated the
development of rocket rams
(weapons).*
**Passenger monorail (power-
driven)**
*cable-drawn system at the Lyon
Exposition, France.*
Solar distillation system
Charles Wilson, UK *engineer at Las
Salinas, Chile, to the designs of J
Harding. It remained in operation for
40 years.*
Woodpulp (sulphite process)
Carl Daniel Ekman *at a mill at
Bergvik, Sweden. By 1875 Ekman
was producing 800 lb of pulp per
annum. The process was soon adopted
in the* USA *and Europe.*

1873
**Barbed wire (mass-production
machine)**
Joseph Glidden *of de Kalb, Illinois,*
USA.
'Burbank potato'
Luther Burbank, US *plant-breeder,
Lunenburg.*
Daguerreotype disc
Pierre Jules César Janssen, *Head of
the Astrophysical Observatory at
Meudon, France – to photograph the
movement of the stars.*
Electric motor (DC)
Zénobe Théophile Gramme, *former
Belgian carpenter. His first, hand-
driven generator was ready in 1871
and two years later he installed
improved models in some French
lighthouses.*
Electromagnetic radiation
James Clerk-Maxwell, *first Professor
of Experimental Physics at
Cambridge,* UK, *and organiser of the
Cavendish Laboratory.*
Fire alarm
Professor Grechi. *A mechanical and
chemical apparatus caused a bell to
ring and a coloured light to flash when
the temperature of the room was
greatly increased.*

Oil (Russian)
Discovered at Baku.
Pyrophone
Frédéric Kastner *of Paris – a chemical harmonicon whose notes were produced by heating glass tubes of various lengths with little flames.*
Silo
Fred L Hatch, US. *He made the first attempt to build a silo above ground in McHenry County, Illinois, USA.*
Typewriter (mass-produced)
Christopher Latham Sholes, *Milwaukee senator and former postmaster/editor, with his friend* Carlos Glidden, *an attorney. Since 1867 Sholes and Glidden had built as many as 40 prototypes, finally managing to sell a much improved machine to the Ilion Arms Manufactory, who called their model 'the Remington' and built the first thousand machines in one year.*

1874
Abecedarium
William Stanley Jevons, UK *economist and logician, while Professor of Logic at Owen's College, Manchester.*
Aircraft (steam-engined)
Félix due Temple, *France. Ramp-launched, it made a very short hop.*
Austria (asteroid)
Johann Palisa, *astronomer, Director of the naval observatory, Pola, Austria. The first of 27 such asteroids he discovered.*
Boyton's life-preserving dress
Captain Paul Boyton, *using Merriman india-rubber as the constructional material for this inflatable suit. Boyton demonstrated his suit on the waterways of the USA, the UK and Europe.*
Cotton wool
Dr Percy, *to purify the air for ventilating the Houses of Parliament, London.*
Electric car
Sir David Salomons *at Tunbridge Wells, Kent, UK. His 1 hp three-*

wheeler was powered from Bunsen cells.
Hydrokimeter (boiler-water-circulator)
Weir Ltd, UK.
Logograph
W H Barlow, *for giving graphic representation to the vibratory motions of the airwaves of speech, like a telegraphic message.*
Merchant ship (triple-expansion engine)
SS *'Propontis', built on the Clyde, UK.*
Pianoforte (sostenuto pedal)
Heinrich *and* Theodore Steinway, US *piano-makers of New York.*
Quadruplex telegraphy
Thomas Alva Edison, US.
Solitons
John Scott Russell, *Scottish engineer.*
Sprinkler system (commercial)
Henry S Parmalee, *New Haven, Connecticut, USA.*
Swinging saloon (steamship)
Sir Henry Bessemer, *with* E J Reed, *naval architect, UK. It was incorporated in the 'Bessemer' cross-Channel steamer as an anti-seasickness device, but was a complete failure.*
Tangent-spoke wheel
James Starley *of Coventry, UK. The spokes were set aslant so that each pair crossed each other at an angle, giving great strength and rigidity. It has been in popular use ever since.*
Tubular steel bridge
Captain James Buchanan Eads, *engineer, and* Charles Shaler Smith, *design consultant, at St Louis, Missouri, over the Mississippi River. It incorporated corrosion-resistant chromium steel in cantilever design, using compressed air caissons.*
Umbrella (curved steel rib frame)
Samuel Fox, *wire-drawer from Deepcar, near Sheffield, UK.*

1875
Canned baked beans
Burnham & Morrill Co. *of Portland,*

Maine, USA, *for their local fishermen.*
Compound locomotive
Anatole Mallet *for the local Bayonne-Biarritz line, France.*
Cube sugar manufacturing process
Eugen Langen, *Cologne, Germany.*
Electric dental drill (battery-operated)
George F Green, *Kalamazoo, Michigan,* USA.
Eosin scarlet dye
Adolf von Baeyer, *Professor of Chemistry at Munich, Germany.*
Gallium
Paul Emile Lecoq de Boisbaudron, *physical chemist, France.*
Magazine rifle
Benjamin Berkeley Hotchkiss, US, *manufacturer resident in France.*
Measurement of wave power
R S Deverell, *Australia.*
Mechanical ship's telegraph
William Chadburn, *Liverpool,* UK, *for transmitting orders from bridge to engine room.*
Milk chocolate
Daniel Peter, *F L Cailler's son-in-law at Vevey, Switzerland.*
Mimeograph
Thomas Alva Edison, US, *while experimenting with paraffin paper for possible use as a telegraphy tape.*

Edison mimeograph

Phenyl-hydrazine
Emil Hermann Fischer, *chemist, Germany.*

Refrigerator freight wagon (railways).
Snooker
Lieutenant Neville Chamberlain, *while serving as subaltern with the Devonshire Regiment at Jubbulpore, India: a variation of 'Black Pool'.*
Solar-collector (axicon-type)
August Mouchot, *French engineer, using truncated cone reflector.*
Tide-calculating machine
William Thomson, *Professor of Natural Philosophy, Glasgow.*
Tinplated food cans (steel).
Xylotechnographica
A P Brophy – *a process for staining wood various colours.*

1876
'Aerotherapy'
Dr Carlo Forlanini, *Pneumatic Health Institute, Milan, Italy.*
Articulating telephone
Alexander Graham Bell, *Scottish-American Professor of Vocal Physiology, Boston University,* USA – *with his assistant* Thomas A Watson. *This invention was also patented by* Elisha Gray *for the Western Telegraph Co., but after Bell.*
Carburettor (surface type)
Gottlieb Daimler *of Cannstatt, Germany, and further modified by* Karl Benz *from 1884 for the 2000 motor cars he sold.*
Carpet sweeper
Melville R Bissell *of Grand Rapids, Michigan,* USA. *As a china shop owner, Bissell suffered terribly from headaches. Convinced the cause was the dusty straw in which the china was packed, Bissell developed a sweeper.*
Cream separator
Gustav de Laval *of Blasenberg. His mechanically whirling arm turbine completely displaced the skimming of cream by hand. His invention was shown at the Royal Show, Kilburn, London, in 1879.*
Crematoria
simultaneously by Julius le Moyne *in*

his own grounds at Washington, Pa, USA, *and by* Dr Pini *of the Milan Cremation Society, Italy.*

Dewey decimal system (library cataloguing)
Melvil Dewey, *acting librarian at Amherst College, Massachusetts,* USA.

Dietheroscope
Giuseppe Luvini *of Tunis, for geodesy and teaching optics.*

Electric candle
Paul Nickolaievitch Jablochkoff, *Russian electro-technologist working in Paris. An electric current was passed through two carbons side by side with a strip of kaolin between them producing a steady, soft, noiseless light — the carbons burning like wax*

Erythrosin and phloxin
Adolf von Baeyer, *organic chemist, Munich, Germany.*

Glaciarum (artificial ice-skating rink)
King's Road, Chelsea, London, constructed by Dr John Gamgee. *Freezing was accomplished by* Raoul Pictet's *liquefaction process and* W E Ludlow's *rotary engine and pump.*

Hygeiopolis (city of health)
Dr Benjamin Ward Richardson. *A company was proposed for its erection, but with no result.*

'Phase rule'
Josiah Willard Gibbs, *mathematical physicist,* US.

Phenylethyl alcohol (synthetic perfume)
B Radziszewski. *It had the scent of roses.*

Plimsoll mark and line
Samuel Plimsoll, MP *for Derby,* UK *and social reformer, 'the sailors' friend'.*

Radiometer
William Crookes FRS, UK.

Rotary-foil hydrofoil craft (steam-engined)
M de Sanderal, *France.*

Stump-jump plough (three-furrow model)
Robert Bowyer Smith, *farmer-*

engineer, Southern Australia. It was called the 'Vixen'.

Tomato ketchup (bottled)
Henry John Heinz & Co., *Pittsburgh, Pennsylvania,* USA.

Twine-binder
John F Appleby, US, *enabling one farm labourer to do the work of eight. It also incorporated a device called a knotting-bill, invented in 1864 by* Jacob Behel *of Illinois,* USA.

1877

Battery (solid depolariser)
Georges Leclanché, *French railway engineer, after ten years' research.*

Cyclometer (distance recorder for 'Ordinary' bicycles)
James C Thompson *of Ealing,* UK.

Dayyum
Vincenz von Kern.

Differential gear
James Starley *of Coventry,* UK, *patented for tricycle transmission to overcome the problem of travelling round bends, when the outer wheel has to run faster than the inner one. This was later adapted for motor vehicles.*

Friedel-Crafts Reaction
Charles Friedel, *chemist and curator of the mineral collections at the School of Mines, Paris, with* James Mason Crafts, US *chemist resident at the school.*

Ilmenium
Hermann.

Liquid oxygen
Louis-Paul Cailetet, *ironmaster of Châtillon-sur-Seine, France, and independently by* Raoul-Pierre Pictet, *physics professor at Geneva University, Switzerland.*

Microphone (dynamic)
simultaneously by Charles Cuttris *with* Jerome Redding, US, *and* E W Siemens, Germany.

Otheoscope
William Crookes, UK, *for studying molecular motion.*

Otto cycle
Nikolaus August Otto, *German*

engineer, who developed a four-stroke internal combustion engine, some ten years after his first mains gas engine, based on the principle proposed by Alphonse Beau de Rochas, France.

Photography (electrically lit).

Resistance welding
Elihu Thomson, *London-born chemistry teacher in Philadelphia,* USA.

Serial photography
Eadweard Muybridge, *with* John D Isaacs, *railway engineer, as commissioned by Californian Governor Leland Stanford, race-horse owner. Using 24 electrically activated cameras lined up beside Palo Alto race-course, their zoogyroscope was a direct forerunner of the modern motion picture.*

Spinthariscope
William Crookes, UK *chemist, for observing ionising radiation.*

Switchboard
Edwin T Holmes. *Installed for Boston's first six telephone subscribers, it served as a telephone system by day*

Telephone exchange

and a burglar-alarm by night.

1878
Cathode ray tube
William Crookes, UK *chemist.*

Electric alternator
Zénobe Gramme *and* Hippolyte Fontaine, *France, and independently by* Werner von Siemens, *Germany.*

Electrically synchronised city clocks
Barraud *and* Lund *for the City of London circuit of 108 clocks.*

Harmonic analyser
William Thomson, Lord Kelvin *and his brother,* Professor James Thomson, *for the Meteorological Office,* UK.

Helical logarithmic calculator
J E Fuller *of New York.*

Lenses (apochromatic objective)
Ernst Abbe, *Professor of Physics and Mathematics at Jena University, together with* Otto Schott, *glass chemist, and* Carl Zeiss, *industrial socialist.*

Megaphone
Thomas Alva Edison, USA, *for use by the deaf.*

Milking machine
L O Colvin, *and manufactured by Albert Durand of Auburn, New York,* USA.

Philippium
Marc Delafontaine, *chemist, Switzerland.*

Phonograph
Thomas Alva Edison and his team of mechanics *at Menlo Park Laboratory, New Jersey,* USA. *The first recording replayed was 'Mary had a little lamb. Its fleece was white as snow.'*

Phonometer (engine powered by human voice)
Thomas Alva Edison *et al,* USA.

Praxinoscope theatre (photographic optical toy)
Emile Reynaud, *France.*

Sailing railway car
C J Bascom *of the Kansas Pacific*

Edison's first phonograph

Railroad. *It was capable of 33 mph and was used for three years to transport repair gangs to pumps, telegraph lines, etc.*

Salvo quad tricycle
James Starley *of Coventry*, UK. *Following the delivery of two vehicles to Queen Victoria, it was re-named the 'Royal Salvo'.*

Solar-powered printing press
Abel Pifre *at the Paris Exhibition.*

Theatre (electrically lit)
the Gaiety Theatre, London.

'Willesden' rot-proof paper and multi-ply green canvas
proofing was by immersion in copper and ammonia, based on the pioneer work of Dr John Scoffern, George Tidcombe *and* Elkanah Haley. *The factory was at Willesden Mills, London.*

Ytterbrium
Jean Charles Galissard de Marignac, *chemist, Geneva, Switzerland.*

1879

Arc lighting system
Edwin James Houston, US *electrical engineer, with his partner,* Elihu Thomson. *A similar system was devised by* Charles F Brush *of Ohio,* USA.

Audiometer (hearing test device)
Professor David Edward Hughes, *London,* UK. *It consisted of a battery of two Leclanché cells, connected with a simple microphone and telephone.*

Benday process
Benjamin Day, *West Hoboken, New Jersey,* USA. *A mechanical method of producing a tinted, shaded or stippled background on a line plate, for the reproduction of maps and illustrations.*

Blasting gelatine (explosive)
Alfred Nobel, *Swedish inventor and manufacturer, and later modified by* Professor Frederick Augustus Abel, UK *War Department chemist.*

'Bordeaux mixture' (fungicide)
M Millardet, *Professor of Botany at Bordeaux, France, to prevent an American mildew disease ruining the grapes in countless vineyards.*

Carbon filament lamp (light bulb)
simultaneously by Joseph Wilson Swan *at Newcastle,* UK, *and* Thomas Alva Edison *at Menlo Park Laboratory, New Jersey,* USA. *Whilst Swan was assisted by high vacua expert Stearn and glass-blower Topham, Edison had a whole team of experts. The major difficulty for both inventors was to find the right material for carbonisation as a filament. Those*

Edison's and Swan's first incandescent lamps

tested included paper, bamboo and human hair.

Cash register
James J Ritty, US *café owner, inspired by a ship's propeller revolution counter. It was built at Dayton, Ohio, but was commercially unsuccessful. Ritty sold his patent for a mere $1000 to what became known as the National Cash Register.*

Coumarin and heliotropin (synthetic perfumes)
having the scent of tonka-beans and heliotrope respectively.

Decipium
Marc Delafontaine, *chemist, Switzerland.*

Domestic refrigerator (steam-pumped)
Karl von Linde, *engineer, Germany.*

Electric streetcar line (temporary)
Werner von Siemens *and* Johann Halske *for the Berlin Industrial Exhibition.*

'Five and Ten Cent Store'
Frank Winfield Woolworth, US *merchant in Lancaster, Pennsylvania, USA.*

'Gabardine'
Thomas Burberry, *draper, of Basingstoke, Hampshire, UK, with Doctor Syme, medical reformer of Edinburgh.*

Holmium
Per Theodor Cleve, *chemist, Uppsala University, Sweden.*

Milk bottles
Echo Farms Dairy Co., *New York, USA.*

Mosandrum
John Lawrence Smith, *chemist, US.*

Norwegium
Telef Dahl.

Rifle magazine box
James Lee, *Scottish watchmaker and US citizen.*

Saccharin
Constantine Fahlberg *and* Professor Ira Remsen *at Johns Hopkins University, Baltimore, USA.*

Samarium
Paul Lecoq de Boisbaudron, *physical chemist, France.*

Scandium
Lars Fredrik Nilson, *chemist, Uppsala University, Sweden.*

Steamship (Siemens-Martin steel-construction)
340 ft 'Pizarro', launched at Napier's Yard, Govan, near Glasgow, UK.

Submarine (steam-powered)
Reverend George W Garrett. *Built at Liverpool, the 'Resurgam' had a two-man crew. Enough steam could be generated to drive her 12 miles after furnaces and chimney had been sealed off for diving. 'Resurgam' was sunk after being towed off Anglesey, UK.*

Temnograph
A M Rymer-Jones, *to plot to any accurate scale a section of ground over which the instrument has travelled.*

Thulium
Per Theodor Cleve, *chemist, Uppsala University, Sweden.*

1880

Audiophone (hearing aid)
R G Rhodes, *Chicago, USA, and improved by M Colladon, Geneva, Switzerland. A thin sheet of hard ebonite rubber or cardboard was placed against the teeth, through which vibrations could be conveyed to the auditory nerve.*

Blowlamp
C R Nyberth *of the Max Sievert Co., Stockholm, Sweden – petrol-burner. The paraffin blowlamp was developed by the* Lindqvist brothers, *also of Sweden.*

Chinese 'alphabet'
Bishop Eligius Cosi *of Canton.*

Cigarette-making machine
James A Bonsack.

Electric clothes iron
France.

Electrically lit town
Wabash, Indiana, illuminated by 4000 candle-power Brush arc lights – in the business quarter.

Electrophone (piped music)
Dr Thaddeus Cahill. *This system whereby subscribers could dial in and listen to performances from theatres and concert halls, via a central switchboard, was also called the 'Telharmonium' or 'Dynamophone'.*
Gadolinium
Jean Charles Galissard de Marignac, *chemist, Geneva, Switzerland.*
Inoculation
Louis Pasteur, *Professor of Chemistry, La Sorbonne, Paris, whilst investigating chicken cholera. Having isolated the germ, he cultivated an attenuated form of it and injected fowls with progressively more virulent suspensions of the culture, so rendering them immune. Following immensely successful inoculation programmes for both anthrax and rabies, the Pasteur Institute was founded in 1888.*
Miniature electric motor
Thomas Alva Edison, *Menlo Park, New Jersey,* USA, *for an electric pen he had designed for producing punctured copying stencils. It was powered by a tiny two-cell battery. 60,000 such miniature motors and pens were manufactured by Western Electric Co.*
Photophone
Alexander Graham Bell *and* Sumner Tainter *of Washington,* USA. *This was a mirror-optical telephone.*
Pre-stressed concrete
P H Jackson, *San Francisco,* USA.
Refrigeration ship
the steamer 'Strathleven' brought to the UK *the first successfully carried refrigerated meat cargo.*
Rubber diaphragm contraceptive (metal-spring rims)
Professor W P J Mensinga.
Solar-powered 2-kW pump
W Adams *for his bungalow at Middle Colaba, Bombay, India.*
Solar spectrum map (infra-red region)
Captain William de Wiveleslie Abney, UK *chemist.*

Submarine-hulled steamship
Sergei Apostolov, *Russia – as a defence against a potential torpedo attack.*

1881
Bolometer
Samuel Pierpont Langley, US *astronomer, to measure radiant heat.*
Compressed-air clock
Victor Popp *of Vienna.*
Cyclostyle pen
David Gestetner, *Hungarian inventor. A spiked-wheel pen which could be used to make a series of perforations in a duplicating waxed stencil.*
Elektromote
Werner von Siemens, *German electrical engineer, and* Johann Halske *– as a trackless car. A line 540 metres long was erected in Halensee and the first demonstration took place in 1882. This was the prototype trolleybus.*
'Farthing-penny' (reversed Ordinary bicycle)
the Smith Machine Co., *New Jersey,* USA. *Called 'the Star', it had the smaller wheel at the front.*
Hydro-electric power station
Messrs Calder & Barrett*'s Central Power Station at Godalming, Surrey,* UK, *for Pullman's leather mill on the River Wey. The principal customer was the Godalming Town Council.*
Interferometer
Albert Abraham Michelson, *ex-*US *Navy academician and physicist. It was used in 1887 for the critical Michelson-Morley experiment.*
Lawn-mower (tricycle-powered)
R Kirkham *of Chicago, Illinois,* USA.
Ocean liner (steel construction)
Canadian Allen Line's *'Buenos Ayrean'.*
Otto dicycle
E C F Otto, *and manufactured by the Birmingham Small Arms Co.,* UK.
Railway (electric)
Werner von Siemens *and* Johann

Halske, *German electrical engineers at Lichterfeld, Berlin – it travelled at 18½ mph.*

Rechargeable battery (lead lattice-work grid)
Camille Faure, *France.*

Telautograph
Elisha Gray, US.

Telephotography
Shelford Bidwell, UK *barrister and physicist – a process of transmitting to a distance images of objects by the agency of electricity and selenium – 'the wirephoto'.*

Télétroscope
Constantin Senlecq, *solicitor of Ardres, Pas-de-Calais, France.*

Tricycle (steam-powered)
Sir Thomas Parkyns *and* William Pateman, UK.

1882

Electric fan (commercial)
Dr Schuyler Skaats Wheeler, *Chief Engineer of the Crocker and Curtis Electric Motor Co. of New York, USA. It was a two-bladed desk fan.*

Electric flat iron
Henry W Seely *of New York,* USA.

Electric mine railway
Werner von Siemens *and* Johann Halske, *electrical engineers at the Zauckeroda coal mine, Saxony.*

Fusil photographique (photographic pistol)
Dr Jules Etienne Marey, *French physiologist. It took twelve consecutive pictures per second. Three years later he had developed his 'chambre chrono-photographique' and reduced exposure time to 1/25,000 of a second to photograph the flight of insects.*

Gas-heated bath tub
Ewart Co's 'General Gordon', Dudley, UK.

Induction coil system
Lucien Gaulard, *France, and* John Gibbs, UK. *They called it a 'secondary generator'.*

Judo
Jigoro Kano, *Japan.*

Oximes
Viktor Meyer, *German chemist at Zurich Polytechnic.*

Skyscraper
William Le Baron Jenny, US *architect, with the Home Insurance Company's ten-storey building in Chicago, Illinois,* USA.

Tuberculosis germ
Robert Koch, *German physician at the Medical School Faculty in Berlin. The following year he discovered the cholera bacillus.*

1883

Automatic temperature control
Albert M Butz, *Minneapolis,* USA.

Automatic vending machine (commercial)
Percival Everitt, *London.*

Diphtheria bacillus
Edwin Krebs *of Zurich, and the following year Friedrich Loeffler succeeded in cultivating the bacillus.*

Dustbin
Eugene Poubelle, *Paris Prefect of Police. It was a galvanised iron portable container.*

Electric blanket
first seen at an exhibition in Vienna.

Electric hair brush
Dr Scott. *'Warranted to cure nervous and bilious headaches, neuralgia and to arrest falling hair and premature greying', it was a complete hoax.*

Gasogen
Emerson Dowson, UK *chemist. It was used by the Crossley brothers to fuel their motor cars.*

Manganese steel
Robert Abbott Hadfield, *Sheffield-born* UK *metallurgist. He also developed silicon steel and stainless steel.*

Narrow-gauge railway (electric)
Magnus Volk. *It ran along the sea-front at Brighton,* UK.

Plastids-from-plastids
Andreas Franz Wilhelm Schimper, *botanist at the Bonn Botanical*

Institute, Germany.

Spark plug (stationary gas engine)
Jean Joseph Etienne Lenoir, *Paris.*

Suspension bridge (long span)
John Augustus Roebling, *German-born US engineer. Brooklyn Bridge, 1600 ft between its towers.*

Thiophene
Viktor Meyer, *German chemist at Zurich Polytechnic.*

1884

Aircraft (steam-engined with triple propellers)
Feodorovitch Mozhaiski, *Russia. Ramp-launched, it made one, very short hop.*

Carburettor (float-feed spray)
Edward Butler *for his 'Petrocycle', UK, then re-invented by both* Wilhelm Maybach *in Germany and* Charles E Duryea *in the USA, some ten years later.*

Cemented bifocals (spectacles).

Evaporated milk
John Mayenberg, *St Louis, USA, and manufactured by Mayenberg's Helvetia Milk Condensing Co., Highland, Illinois.*

'Farthing-Penny' (steam-powered bicycle)
Lucius D Copeland *of Phoenix, Arizona, USA – a converted 'Star' bicycle.*

Fountain pen
Lewis Edson Waterman, *US insurance salesman. Its capillary feed system made for a much freer flowing reservoir pen than those produced since the early 1800s.*

Greenwich Mean Time.

Kangaroo (bicycle)
E C F Otto *and* J Wallis, *UK, and manufactured by Hillman, Herbert & Cooper of Coventry.*

Local anaesthetic (cocaine)
K Köller, *Austrian ophthalmologist.*

Machine-gun (water-cooled and belt-loaded)
Hiram Stevens Maxim, *US inventor of*

Huguenot descent, who had settled in the UK (see overleaf).

Nipkow scanning disc
Paul Nipkow, *German engineer.*

Oil well (Sumatra)
Aeilko Jans Zijlker, *Dutch engineer.*

Phillips dipping entry wing
Horatio Phillips, UK, *who discovered that an aerofoil will provide lift if the upper surface is deeply curved and the lower surface is shallowly curved.*

Rollercoaster
L N Thompson, *founder of Luna Park, Coney Island, New York, USA. It was called the 'Switchback'.*

Steam-powered turbine engine
the Hon Charles Algernon Parsons, UK *engineer at a dynamo-manufacturing firm, Gateshead, County Durham.*

Synchronous multiplex telegraphy
Patrick Delany, *New York.*

Test for bacteria
Hans Christian Joachim Gram, *Danish bacteriologist.*

Transformer (practical commercial)
Max Deri, Otto Blathy *and* Karl Zipernowsky *of Hungary; also* Lucien Gaulard *of France.* William Stanley, *electrical engineer for the Westinghouse Company, US, invented his version of the transformer less than a year later.*

Washing machine (octagonally shaped and hand-rotated)
Thomas Bradford, UK *engineer.*

1885

Ammonium picrate (explosive)
Eugène Turpin, *France. Called 'Mélinite'.*

Commercial plywood factory
Christian Luther, *in Revel, Estonia, manufacturing 3-ply chairseats.*

Electric toothbrush
Dr Scott, US.

'Flying Bedstead'
W O Ayers, US. *Hand and foot pedals and compressed air motors drove seven*

Maxim beside his machine gun, (1884)

propellers – *six for lift and one for propulsion. It never flew as it was never built.*

'Golden syrup' (canned treacle) Abram Lyle III, *Scottish sugar refiner, with his sons, at Plaistow, East London.*

Holiday caravan (custom-built) Dr Gordon Stables RN, *and built by the Bristol Wagon Co. Fully equipped and horsedrawn, 'The Wanderer' made a 1300-mile tour of the UK with Dr Stables's valet, Foley, travelling in advance on a tricycle to warn other road-users of its approach.*

Humber tandem tricycle Thomas Humber *and* T H Lambert, *Beeston, Yorkshire, UK.*

Incandescent fabric gas mantle Karl Auer, *Vienna-born chemist and former pupil of Robert Bunsen.*

Moray car G A Thrupp, *London coachbuilder.*

Motor-bicycle (petrol-engined) Gottlieb Daimler *and* Wilhelm Maybach *at Cannstatt, Germany. It was the first to incorporate 'twistgrip' handlebar acceleration.*

Motor car (petrol-engined) *simultaneously by* Gottlieb Daimler *with* Wilhelm Maybach, *engineers at Cannstatt; and* Karl Friedrich Benz, *engineer at the Rheinisch Gasmotorenfabrik, Mannheim, Germany.*

Neodymium Karl Auer, *Vienna-born chemist.*

Petrol pump Sylvanus F Bowser, *Fort Wayne, Indiana,* USA, *to enable local storekeeper Jake Gumper to supply his customers with kerosene lamp oil in given quantities.*

Praseodymium Karl Auer, *Vienna-born chemist.*

Safety bicycle John Kemp Starley, *whose Rover*

Mark III with equal-sized wire-spoked wheels and diamond frame was the final design which ousted the highly unstable Ordinary (penny-farthing) of the 1870s. That Starley's Rover had its crank and pedals in the centre with a chain-drive to the back wheel was thanks to H J Lawson's 'bicyclette' of 1879. The only feature it still lacked was an efficient tyre.

1886

Aluminium (by electrolysis)
simultaneously by Paul Louis Toussaint Héroult, aged 23, in France, and by Charles Martin Hall, aged 22, of Oberlin, Ohio, USA, working without collaboration. Both students made aluminium by the electrolysis of a solution of aluminium oxide in fused cryolite. Soon after, a water-powered aluminium factory was formed at Neuhausen on the Swiss-German frontier, with Martin Kiliani as its chief engineer. In 1889, the first 40 tons of pure aluminium were electrolysed.

Automatic air brake (railways)
George Westinghouse, Pittsburgh, Pennsylvania, USA.

Boat (internal combustion engine)
Gottlieb Daimler of Cannstatt, Germany. Secretive dawn trials were held along the local River Neckar.

Canal rays
Eugen Goldstein, German physicist

Celesta
Auguste Mustel, Paris instrument-maker, as a keyboard glockenspiel.

Coca Cola
Dr John Pemberton, Atlanta, Georgia; marketed as 'Esteemed Brain Tonic and Intellectual Beverage'.

Coin-operated ticket machine
James Longley, and installed at Leamington Athletic Ground, Leeds, UK.

Comptometer (key-driven calculating machine)
Dorr Eugene Felt, Chicago, Illinois, USA.

Diapason clock
Louis Breguet, French clockmaker.

Dinner jacket
Griswold Lorillard, at the Autumn Ball held at the Tuxedo Park Country Club, New York. It became known as 'Lorillard's Tuxedo'.

Dysprosium
Paul Emile Lecoq de Boisbaudron, France.

Electric light bulb (bayonet cap)
Edison & Swan United Electric Light Co., Newcastle-on-Tyne, UK.

Germanium
Clemens Alexander Winkler, chemist, Freiburg, Germany.

Graphophone
Chichester Bell and Charles Sumner Tainter, US scientists, using a wax cylinder and a gouge-shaped sapphire stylus.

Hatchet planimeter
Prytz.

Linotype machine
Ottmar Mergenthaler, German-born engineer working in Baltimore, Maryland, USA. Following eight years of prototypes, Mergenthaler's 'blower machine' enabled type to be set mechanically, using a keyboard. The first 30 machines were installed in the 'New York Tribune' building.

Mechanical potato digger.

Motorboat (paraffin vapour)
William and Samuel Priestman of Hull, UK.

Photographic hat
Lüders Brothers, Gorlitz, Germany.

Sound signal cyclometer
Messrs C V Boys and M D Rucker.

Submarine (electric)
Lieutenant Isaac Peral, Spanish officer. The Peral was powered by 480 storage accumulators powering two 30 hp motors turning two propellers.

Tintometer
Joseph William Lovibond, brewer, maltster and spirit merchant of Salisbury, UK.

1887

Automatic page-turner (piano music)
Augustin Lajarrige, *mechanic of Marseilles, France.*

Collapsible pocket umbrella.

Contact lenses
Dr Eugen A Frick *of Zurich.*

Electric elevator
Werner von Siemens *and* Johann Halske, *electrical engineers, for the Industrial Exposition at Mannheim, Germany.*

Electric heater
Dr W Leigh Burton. *Commercially manufactured by the Burton Electric Co. of Richmond, Virginia,* USA, *it was first used on electric tramcars.*

Electrocardiogram
Augustus Désiré Waller, *Parisian-born physiologist at London University.*

Electro-magnetic waves
Heinrich Rudolf Hertz, *while physics professor at the Karlsruhe Polytechnic and aged only 30. Hertz (Hz) remains the unit of measurement.*

Esperanto (auxiliary language)
Dr Lazarus Ludwig Zamenhof, *Polish oculist.*

Gas meter (coin-in-the-slot)
R W Brownhill *of Birmingham,* UK. *Only five years later did the South Metropolitan Gas Co. agree to instal 100 experimental meters.*

Gold extraction (cyanide process)
John Stewart Macarthur, *and the* Forrest brothers.

Gramophone record
Emile Berliner, *German immigrant living in Washington* DC, USA. *He also introduced the copying of flat disc recordings by making a wax disc from which a 'negative' metal matrix was made for producing endless positives. Although the first commercial manufacture was by Kammerer and Rheinhardt of Waltershausen, Germany, greater success was achieved by the Gramophone Co. formed eleven*

years later.

Hydrazine
Theodor Curtius, *German organic chemist.*

Impact turbine (steam)
Carl Gustav Patrik de Laval, *Swedish engineer.*

Mach scale and angle
Professor Ernst Mach *of the Physics Faculties of Prague and Vienna, while studying supersonic projectiles and the flow of gases.*

Malted milk
William Horlick, *Racine, Wisconsin,* USA.

'Parlour' hansom cab
Joseph Parlour.

Phonograph doll
Thomas Alva Edison, *Menlo Park, New Jersey,* USA.

Pocket typewriter
Messrs Dobson & Wyn, *London.*

Rhodamine
Johan Friedrich Adolf von Baeyer, *Professor of Chemistry, Munich, Germany.*

Rotary aero engine (experimental)
Lawrence Hargrave, *Australia.*

1888

Ballistite
Alfred Nobel, *Swedish inventor and manufacturer.*

Cellulose film
John Carbutt, UK *photographer working in the* USA. *Carbutt persuaded a celluloid manufacturer to produce sufficiently thin sheets, coated with gelatine emulsion, for him to roll through a camera instead of the cumbersome glass-plate negative system.*

Compressed air tramcar
tested by the London Steam Tramway Company on its Caledonian route between King's Cross and the Holloway Road. It was found unable to climb hills.

Crude oil refining process
Hermann Frasch, *German-born*

Alfred Nobel, as depicted on the Nobel Prize medal

pharmacist, employed by the Standard Oil Co. of Ohio, USA.

Electric induction motor (AC)
simultaneously by Nikola Tesla, Croatian-born engineer, at his laboratory in New York and by Professor Galileo Ferraris in Turin, Italy. Both men demonstrated the practical applications of alternating electrical current.

Electric switch socket
Sir David Salomons, Tunbridge Wells, Kent, UK.

Electric tramway (regular service)
Frank J Sprague, US. The route, in Richmond, Virginia, was 12 miles long.

Electric vehicle disc brakes
Elmer Ambrose Sperry, inventor, Chicago, Illinois, USA.

Flexible roll-film camera
George Eastman, dry-plate manufacturer from Rochester, New York, USA. Called the 'Kodak', it was given the slogan 'You press the button, we do the rest.'

Flexible, unbreakable beer glass
G W A Kahlbaum, Swiss chemist, using methylacrylate polymers.

Monorail
Charles Lartigue. The 9-mile line ran between Listowel and Balybunion,

Ireland.

Monotype
Tolbert Lanston, a Washington civil servant. The type-forming and composing machine was in production by 1897.

Motor tricycle (production model)
Karl Benz, as marketed by Roger in Paris.

Phonograph (commercial model)
Thomas Alva Edison, Menlo Park, New Jersey, USA. It had a clockwork motor and wax cylinders.

Pneumatic bicycle tyre
John Boyd Dunlop, prosperous Scottish veterinary surgeon living in Belfast, to help his ten-year-old son, Johnnie, win a race over the bumpy cobblestones of the city on his Edlin Quadrant tricycle. Hosepipe sections were filled with air, largely on the advice of the Dunlops' family doctor, Sir John Fagan, from experience he had gained with medical air beds and inflatable cushions. Following a patent, Dunlop went into partnership with Harvey DuCros, Irish industrialist, to form a manufacturing company.

Recording adding machine (full keyboard)
William S Burroughs, American Arithmometer Co., St Louis, Missouri, USA.

Revolving door
Theophilus van Kannel, USA, for use in the new skyscrapers, where air pressure outside and suction from elevators inside made conventional doors difficult to open.

Solar thermocouple
E Weston, who designed a machine whereby the solar energy, instead of focusing on mirror or plates, heated up a thermocouple so as to create a flow of electricity.

Submarine (battery-electric motor)
Gustav Zédé's 'Gymnote', built to

plans of H Dupuy de Lôme, France.

1889

Active molecules
Svante August Arrhenius, *Swedish scientist.*

Agricultural tractor (petrol-engined)
Charter Engine Co. *of Chicago, Illinois. It was used on wheat farms in the Dakotas,* USA.

Arithmographe (pocket calculator)
Troncète, *France.*

Artificial silk
Hilaire, Comte de Chardonnet, *French chemist. His silk was very inflammable so that his workmen called it 'mother-in-law' silk, since the present of a dress made of it was an effective means of disposing of a troublesome relative.*

Automatic telephone exchange
Almon B Strowger, *Kansas City, undertaker.*

'Bird Flight as the Basis of Aviation' (book)
Otto Lilienthal, *German civil engineer and man-carrying-glider innovator.*

Celluloid roll film
Henry Reichenbach, *at George Eastman's Kodak factory, Rochester, New York.*

Cordite
James Dewar *and* Frederick Augustus Abel. *Abel was chemist to the British War Department and Ordnance Committees.*

Cotton-picker
Angus Campbell, *pattern-maker from Chicago,* USA. *For the next 30 years, with financial backing from Theodore H Price, a wealthy cotton dealer, Campbell continued to develop and improve his spindle-type picker including equipping it with a petrol engine.*

Data-processing computer
Dr Herman Hollerith *of New York, for the* US *Census Bureau. It used*

punched cards containing information submitted by respondents to the census questionnaire. Seven years later, Hollerith established his Tabulating Machine Co.

Dish-washing machine (power model commercially produced)
Mrs W A Cockran *of Shelbyville, Indiana,* USA, *after ten years of prototypes.*

Eiffel Tower
Gustave Eiffel, *French engineer, based on experience gained whilst building iron bridges at Oporto and Garabit. In Eiffel's office, 40 draughtsmen turned out 5000 drawings for the 985-ft high tower. The sections were prefabricated at his factory. The completed tower cost £260,000.*

Electric oven (experimental)
Hotel Bernina in Samaden, Switzerland, which had its own electric power supply, generated from a dynamo, driven by a waterfall.

Electric tricycle
John Kemp Starley *of Coventry,* UK.

Fruit machine
Charles Frey, *German immigrant at San Francisco,* USA. *It was called the 'Liberty Bell'.*

Kinetoscope (peep-show machine)
Thomas Alva Edison *and his assistant,* William Kennedy Laurie Dickson, *at Menlo Park Laboratory, New Jersey,* USA. *It used 158 glass plates, but later progressed to Eastman's and Goodwin's roll film.*

Life boat (steam-powered)
the British water-jet-propelled 'Duke of Northumberland'.

Lysine (amino acid)
Edmund Drechsel, *physiological chemist at Carl Ludwig's Institute, Leipzig.*

Northrup loom
James H Northrup, *English mechanic working for George Draper & Sons of Hopedale, Massachusetts,* USA. *The first commercially produced machine went into action at Seaconnet Mill,*

Fall River.

Plankton
Viktor Hensen, *Professor of Physiology, Kiel University, Germany.*

Pneumatic yacht (gas-jet-propulsion)
'Eureka' of Brooklyn, New York, USA.

Recording adding machine
Felt & Tarrant Manufacturing Co. *of Chicago,* USA. *It was called the 'Comptograph'.*

Skyscraper (riveted skeleton)
William Holabird *and* Martin Roche, US *architects, with the Tacoma Building, Chicago,* USA.

Table-tennis
James Gibb, *engineer and distinguished amateur athlete of Croydon, Surrey,* UK. *It was manufactured nine years later by John Jaques & Son Ltd and marketed by Hamley Brothers of Regent Street, London, first as 'Gossima' and soon after as 'Ping Pong'.*

Telephone call-box (coin-operated)
William Gray. *Following the installation of a prototype at Hartford Bank, Hartford, Connecticut,* USA, *the Gray Telephone Pay Station Co. was formed to rent out coin-operated telephones to store-keepers.*

Water closet (washdown flushing)
David Bostel, *plumber, Brighton,* UK.

1890

Aeroplane (man-carrying)
Clément Ader, *whose steam-engined 'Eole' 'hopped' 50 metres at Armainvilliers, France, with its inventor at the controls.*

All-steel oil rig.

Aluminium saucepan
Henry W Avery, *Cleveland, Ohio,* USA.

Babcock test
Professor S M Babcock *of the University of Wisconsin,* USA, *by which at a glance a farmer could see the*

Clement Ader's 'Eole'

percentage of butter fat in milk or cream on a graduated container. Although this test revolutionised dairy manufacture and the breeding of dairy animals, Professor Babcock refused to patent this or any of his other inventions, preferring that the world might benefit from them.

Compression-ignition engine
Herbert Akroyd Stuart, *and produced by Richard Hornsby & Sons of Grantham, Lincolnshire,* UK, *two years later. It was called the 'hot bulb' engine.*

Cuprammonium rayon
Louis Despeissis, *France.*

Diphtheria antitoxin
Emil Adolf von Behring, *German bacteriologist, and* Shibasaburo Kitasato, *Japanese investigator at Robert Koch's Institute for Infectious Diseases, Berlin. It was placed on the market in 1892.*

Electric bicycle
G P Hachenberger, *Austin, Texas,* USA. *It was to be powered and supported by electric telegraph wires.*

Electric chair
Harold P Brown *working with* Thomas Alva Edison's *chief electrician,* Dr E A Kennelly. *Following experiments with many animals, William Kemmler, convicted murderer, was executed at Auburn Prison, New York,* USA. *He took 8 minutes to die.*

Electric-generator windmill
P la Cour *in Denmark, who mounted patent sails and twin fantails on a steel tower.*

Elevated and surface railroad system
T C Clarke, *consulting engineer, New York,* USA.

High-frequency generator
Elihu Thomson, US *engineer and inventor, with* Edward James Houston.

Ice-cream sundae
Jed Smithson, *Wisconsin,* USA.

Kinematograph camera and projector
William Friese-Greene, *Bristol-born photographer and self-taught scientist, working in Holborn, London. His fourth camera, able to take pictures on modified celluloid film at 50 frames per second by means of parallel sprocket holes, was publicly demonstrated in Chester Town Hall during this year.*

Machine-gun (gas-operated)
Baron Adolf Odkolek von Augezd, *Austrian army officer.*

Pure nickel process
Monde, Langer *and* Quincke.

Radio conducteur (coherer)
Edouard Branly, *Physics Professor at the Catholic University, Paris.*

'Schizophone'
Captain Louis de Place, *France, to detect internal cracks or holes in metals.*

Underground electric railway
London. *The City and South London Railway ran from Arthur Street to Stockwell, a distance of 3 miles, passing under the River Thames in separate tunnels, with one class and one fare, turnstiles instead of tickets, and hydraulic lifts or winding stairs. The initial idea of traction by an endless cable was abandoned in favour of electric power as the motive force.*

Wells hive (two-queen system of bee-keeping)
George Wells, *bee-keeper, Kent,* UK.

1891
Aluminium boat
Escher, Wyss & Co. *of Zurich, Switzerland. Even its smoke stack and rigging were of aluminium. It was powered by a 2 hp naphtha engine, had a speed of 10 km per hour, weighed only 1000 lb and carried eight passengers. Soon after, the same company built the 43 ft naphtha-engined yacht, 'Mignon', whose machinery was of aluminium except for the cranks and shafting.*

Bazeries cylindrical cipher device
Captain Etienne Bazeriȩs, *French cryptologist.*

Crankshaft compression phase (two-stroke engine)
Joseph Day.

Electric oven (commercially manufactured)
Carpenter Electric Heating Manufacturing Co., *of St Paul, Minnesota,* USA.

Electric torch
Bristol Electric Lamp Co., *whose 2-candle-power bull's eye lantern was used by ticket inspectors of the Bristol General Omnibus Co.,* UK.

Motor car (electric)
William Morrison, US *engineer. The vehicle was powered by 24 electric storage batteries placed under the seats giving it a total running time of thirteen hours with a passenger load of twelve people.*

Musk xyol and musk ambrette (synthetic perfumes)
A Baur.

Pneumatic typewriter
Marshal A Weir, *London.*

Premature baby incubator
Dr Alexandre Lion *of Nice, France. 137 out of 185 children reared in the incubators at Nice during the first three years survived infancy. Dr Lion introduced his 'Couveuses' to London in 1897.*

Silicon carbide
Eduard Goodrich Acheson, US. *By 1893 it had come into production as the abrasive, carborundum.*

Spectroheliograph
George Ellery Hale, US *astronomer at*

Kenwood Observatory, Chicago, Illinois, USA.
Telephoto lens (camera)
Thomas R Dallmeyer, *lens manufacturer,* UK. *Also* Dr Miethe *of Germany and* A Dubosq *of France.*
Traveller's cheque
American Express Co.
Zip fastener
Whitcombe L Judson, US *mechanical engineer. It was first exhibited at the Chicago Exposition two years later. Initially the Universal Fastener Co. produced the slide fastener by hand, but following a re-design by Judson in 1905, the Automatic Hook and Eye Co. of Hoboken, New Jersey began machine-producing them under the trade name of 'C-Curity'.*

1892
Cellulose-fibre manufacture (viscose process)
Charles F Cross, UK *consulting chemist, and his partner* Edward John Bevan. *It went into commercial manufacture by* C H Stearn *and* Topham *as an artificial silk.*
Cholera vaccine
Waldemar Mordecai Wolff Haffkine, *Russian bacteriologist and physician working in India.*
Crown top and cast-iron bottle opener
William Painter, *Baltimore, Massachusetts,* USA.
Crum Brown's Rule
Alexander Crum Brown, *Professor of Chemistry, Edinburgh,* UK.
Diesel engine
Rudolf Diesel, *German engineer, subsidised by Krupps, after fourteen years' research on a compression-ignition engine. After a further six years' developmental work at Augsburg, manufacturing rights were acquired by Russia, Sweden and France.*
Electric fire (domestic)
Colonel Rookes Evelyn Bell Crompton *and* Herbert John

Dowsing, UK.
Iodoso compounds
Viktor Meyer, *German chemist at Heidelberg.*
Magnetic controller (for electric trolley cars)
Elihu Thomson, *Swampscott, Massachusetts,* USA.
Mega-globe
Alberto di Palicio, *Spanish architect, to be erected on San Salvadore to commemorate the 400th anniversary of Christopher Columbus's discovery of America. The globe was to be 100 ft in diameter, on a concrete base 250 ft high and topped by a huge statue of Columbus. Its concrete base was to contain museums, lecture halls, public meeting places and a planetarium. It was never built.*
Motorcycle (rotary-engined)
Félix Theodore Millet, *France.*
Outboard engine (petrol)
Alfred Seguin *of France, at the SNG garage in Geneva, Switzerland. It was called the 'Motogodille'.*
Phagocytes (cells)
Ilya Mechnikov, *Russian biologist, working at the École Normale, Paris.*
Portable typewriter
George C Blickensderfer, US. *Called the 'Featherweight Blick', it was based on an aluminium chassis.*
Toothpaste tube (collapsible metal)
Dr Washington Sheffield, *dentist of New London, Connecticut,* USA, *and later manufactured by his Sheffield Tube Corporation. Beecham's Tooth Paste was retailed in tubes the same year.*
Vacuum flask
Sir James Dewar, *Cambridge,* UK, *for laboratory work. It was commercialised in 1904 by Reinhold Burger, Germany, as the 'Thermos' flask.*
Virus
Dmitri Iosifovich Ivanovsky, *botanist, St Petersburg, Russia.*

1893

Aerograph air-brush
Charles L Burdick, US *artist working in London, who founded the Fountain Brush Co. in Clerkenwell Green.*

Breakfast cereal
Henry D Perky, *lawyer of Denver, Colorado*, USA – *'Shredded Wheat'.*

Electric toaster
Crompton & Co., UK.

Ionone (synthetic perfume)
Johann C W F Tiemann, *chemistry professor at the University of Berlin, Germany, with* P Kruger – *having the scent of violets.*

Kinetographic theatre (film studio)
W K L Dickson *at Menlo Park, New Jersey*, USA – *the studio revolved to follow the sunlight.*

Kirby grip hairpin
Hindes.

Language-teaching course (phonographs)
Dr Richard Rosenthal *of the International Institute of Languages, London and New York, who produced 50 wax cylinders with accompanying text.*

Lawn-mower (steam-powered)
James Sumner *of the Lancashire Steam Motor Co.*, UK. *150 1½-ton machines were sold, many for large country estates.*

'Model electric kitchen'
Chicago World's Fair, including both electric saucepan and electric kettle.

Photo-electric cell
Julius Elster *and* Hans F Geitel. *Geitel was Professor of Physics at the Brunswick Technical Institute, Germany.*

Solar-electric engine
M L Severy, *whose engine operated in conjunction with wet storage batteries to enable the user to have 24-hour electric power.*

1894

Aircraft (twin steam-engined)
Sir Hiram Maxim. *Unmanned, powered by two 180 hp engines, with wind-tunnel-tested aerofoils, it failed to fly.*

Argon gas
John William Strutt, UK *physicist, and* William Ramsay, *Scottish chemist at University College, London.*

Chromatic harp
Gustave Lyon *of Paris.*

Electric arc furnace
Ferdinand Henri Moissan, *Professor of Chemistry, University of Paris, France.*

Employee's time-recorder (clock-in/clock-out)
Daniel M Cooper, US. *The operation of a lever printed the time on a card divided by lines into seven equal spaces for the days of the week.*

Escalator
Jesse W Reno, *New York entrepreneur, as a novelty ride on Coney Island Pier, New York.*
Another designer of 'inclined elevators' was Charles Seeberger. *Before long escalators had been installed in department stores and railway stations.*

Helium
William Ramsay, *Professor of Chemistry, University College, London.*

Lorry (petrol-engined)
René Panhard *and* Émile Levassor, *coachbuilders, D'Ivry, Paris.*

Meteorograph-carrying kite
W A Eddy, *above the* US *Blue Hill Weather Observatory, Massachusetts.*

Motor barge
Delamare-Debouteville *and* Malandrin. *'L'Idée' went on successful trials on the Tancarville Canal, Le Havre, France.*

Motorcycle (pneumatic-tyred)
Heinrich *and* Wilhelm Hildebrand *and* Alois Wolfmüller, *German engineers.*

Wireless telegraphy
Guglielmo Marconi, *twenty-year-old*

Italian at Villa Griffone, his family home outside Bologna. By *1897*, the Marconis had formed their Wireless Telegraph and Signal Co. in London.

1895

Aerial
Professor Aleksandr Stepanovich Popov *of Kronstadt.*

Amphibious passenger train (steam-engined)
Magnell. *Built by Ljunggreen of Christianbad, Sweden, it ran on water rails between Furum and Fredericksdal, near Copenhagen, Denmark. By 1897 'Swan' had carried 40,000 passengers.*

Baked beans in tomato sauce (canned)
H J Heinz Co., *Pittsburgh, Pennsylvania,* USA.

Chiropractic (spinal re-adjustment)
Daniel David Palmer, *osteopath,* USA.

Ciné film show
Louis *and* Auguste Lumière, *French brothers. A sequence of films was shown in a Parisian café: a train entering a station, a rowing boat leaving a harbour and workers coming out of the Lumière factory at Lyons.*

'Daddy Longlegs'
Magnus Volk – *an electrically-driven pier-boat on stilts.*

Electric hand drill
Wilhelm Fein, *Stuttgart, Germany.*

Epicyclic gearbox
Dr Frederick William Lanchester, *engineer, Birmingham,* UK.

Fish paste (in jars)
Charles Shipham, *high-class pork butcher, Chichester, Sussex,* UK.

Flaked breakfast cereal
Dr John Kellogg *of the Battle Creek Sanatorium, Michigan,* USA, *with 'Granose Flakes'. 'Corn Flakes' were first manufactured three years later by Kellogg's brother William and marketed by Sanitas Food of Battle Creek.*

Glass-blowing machine
Michael J Owens *of Toledo, Ohio,* USA. *His production model was ready by 1904.*

'Invar' (nickel-steel alloy)
Charles Edouard Guillaume, *Swiss metrologist, International Bureau of Weights and Measures, Sèvres, near Paris.*

'Latham Loop' (motion picture camera device)
Enoch J Rector, US, *working for the Latham brothers – to prevent the snapping of film. Exactly the same device was invented the same year by Thomas Armat for his Vitascope 'Beater' Movement film projector.*

Microtones (musical notation)
Julián Carillo, *Mexican composer.*

Milking machine (pulsator model)
Dr Alexander Shields *of Glasgow,* UK, *with the Thistle Mechanical Milking Machine.*

Motor ambulance
a Daimler-engined Panhard and Levassor.

Motor bus (petrol-engined)
Netephener Omnibus Co.'s *5 hp Benz single-decker enclosed landau, for a 15-km route in the North Rhineland, Germany.*

Motor cycle (four-cylinder)
Captain Capel Holden, UK *civil engineer.*

Motorcycle (quantity-produced)
Marquis Albert de Dion *and* Georges Bouton, *France.*

Photographic portrait machine
Ferrer.

Photographic typesetting
William Friese-Greene, *kinematography inventor of London, who took out a patent, although his system was never built for commercial use.*

Pneumatic-tyred automobile
André *and* Edouard Michelin, *brothers, on a Peugeot during the Paris-Bordeaux-Paris race. They were forced to retire after innumerable*

*punctures and bursts, despite carrying
22 spare inner tubes on board.*

Rotogravure
Karl Klietsch *in Vienna.*

Talking watch (phonographic)
Sivan, *watchmaker, Geneva,
Switzerland.*

Time machine
Herbert George Wells, *fictional
writer,* UK – *published as a short story,
still to be built as a practical reality.*

Triplet hydrocycle
Messrs Keen, Marriott & Cooper *of
Holborn Viaduct, London. It was
successfully tested on the River
Thames.*

**Tse-tse fly (as sleeping sickness
carrier)**
David Bruce, *Australian-born
physician, Royal Army Medical
College, Millbank,* UK.

X-rays
Wilhelm Könrad von Röntgen,
*Director of the Physical Institute,
Würzburg, Germany.*

One of Röntgen's cathode-ray tubes

1896
Armoured car
E J Pennington, US *at Motor Mills,
Coventry,* UK. *The four-man, 16 hp
Pennington Military Autocar had
Harveyised steel armour and twin
machine guns.*

**Automatic oyster-picking
machine**
modified diving bell.

**Automatic rotary stencil
machine**
Lowe *in the* USA *and* Ellams *in the* UK.

Blériot acetylene headlamps
Louis Blériot, *Paris.*

**Centrifugal-type speedometer
(motor car)**
Edward Prew, *schoolmaster,
Yorkshire,* UK.

Dial telephones
*Private automatic exchange, City
Hall, Milwaukee,* USA.

Direct top gear (motor car)
Dr Frederick William Lanchester,
engineer, Birmingham, UK.

Electrons
Joseph John Thomson, *Cavendish
Professor of Experimental Physics,
Trinity College, Cambridge,* UK.

Europium
Eugène Anatole Demarçay, *French
chemist.*

Four-cylinder motor car engine
René Panhard *and* Émile Levassor,
France.

Histidine (amino acid)
Albrecht Kossel, *professor of
biochemistry at Heidelberg, and* Sven
Anders Hedin, *professor at Uppsala,
Sweden.*

Ice-cream cones
Italo Marcioni, *Italian immigrant to
New Jersey,* USA.

Ion-radiated food (theory of)
both Wilhelm Könrad von Röntgen,
German physicist, and Antoine Henri
Becquerel, *French physicist.*

Mechanically-tipping hat
James Boyle *of Washington,* USA.

Motor car (production model)
René Panhard *and* Émile Levassor,
former coachbuilders of Paris.

Periscope (submarine device)
Gustav Zedé, *France, for use on the
'Morse'.*

Portable propeller pitchometer
J Chapman *and* J Hunter, UK.

Rear-view vehicle mirror
John William Cockerill, *Captain
Surgeon, Army medical staff, London.*

Septic tank (sewage treatment)
Donald Cameron *of Exeter,* UK. *The
effluent from the septic tank was
treated on contact-beds as developed*

several years before by W J Dibden.
Sphygmomanometer
Dr Scipione Riva-Rocci, *Italy*.
Steamroller-ship
Ernest Bazin, *St Denis, France*. *180
ft, with a carrying capacity of 300
passengers, it would roll over the
waves. Following satisfactory trials
under Vice-Admiral Coulembeaud,
the project was abandoned.*
Visible typewriter
Herman L Wagner, US. *It was called
the 'Underwood Number One'. With
previous machines the typist was
unable to see what he/she was typing
from word to word.*

1897
Agrimotor (petrol-engined)
Messrs Hornsby *with their Hornsby-
Ackroyd Patent Safety Locomotive of
18 bhp for working threshing
machinery, road haulage and other
tasks.*
Bismuth effect (radiology)
Walter Cannon, *physiology student at
Harvard University*, USA.
Cathode-ray tube
Ferdinand Braun, *Professor of
Physics, Strasbourg, Germany*.
Cell-free fermentation
Eduard Buchner, *chemist, Tübingen,
Germany*.
Decumtyplet bicycle
Waltham Manufacturing Company,

USA, *for ten riders as a pacing machine.
Soon after, a Quindicuplet was
constructed for fifteen riders.*
Digestion physiology
Ivan Petrovic Pavlov, *Professor of
Physiology at the Military Medical
Academy, St Petersburg, Russia*.
Electric omnibus
one tested in London.
**Filaria (malaria-carrying
parasite)**
Ronald Ross, *officer in the Indian
Medical Service, and* C Finlay *in
Cuba. Work completed by* Patrick
Manson, *Scottish doctor.*
Foot accelerator
Dr Frederick William Lanchester,
Birmingham, UK.
**Galalith and Erinoid (casein
plastics)**
Krische *and* Spitteler, *to meet the
unusual demand in Germany for white
blackboards made from milk.*
Hydrofoil (steam-powered)
the Comte de Lambert, *working with
Englishman*, Horatio Phillips, *on the
River Seine, France*.
Oscillograph
William du Bois Duddell, UK
engineer.
Pianola
Edwin S Votey, *Farrand & Votey
Organ Co., Detroit, Michigan*, USA.
*Perforated paper-roll-automata-
piano.*

Plasticine
manufactured by William Harbutt,
Bath, UK.
Scooter
Walter Lines, *fifteen-year-old London schoolboy who, when he grew up, founded the Triang Toy Company.*
Simms-Bosch low tension magneto
Frederick R Simms, UK; *manufactured in Stuttgart by Robert Bosch.*
Taximeter motor cab
Friedrich Greiner, *Stuttgart, Germany.*
Turbine-driven steam yacht
the Hon. Charles Algernon Parsons, *Newcastle-upon-Tyne*, UK. *The 100-ft 'Turbinia' achieved the unheard-of speed of 34.5 knots.*
Umbrella boat
Percy Pilcher *and* W G Wilson, UK. *They rigged up a 'cyclone sail' for a 17-ft boat and tested it on the Solent. Her 360 sq ft of umbrella gave her a stable swiftness, while only 200 ft of conventional canvas proved too much.*
Wildlife-preserving park
Umfolozi Game Park, Natal, South Africa.
Worm gear
Dr Frederick William Lanchester, *engineer, Birmingham*, UK.

1898
Anti-neuritic Vitamin B
Christiaan Eijkman *and his assistant* G Grijns *at Batavia in the Netherlands Indies (present-day Indonesia).*
'Bellowing telephone' (loudspeaker)
Sir Oliver Lodge, *Professor of Physics, University College, Liverpool*, UK.
Direct-drive motor car
Louis Renault, *France. An adapted De Dion Bouton tricycle, it led to a production model.*
Electric torch (tubular)
Electric and Novelty Manufacturing Co. *of New York – later the Ever-Ready Company.*
Garden cities
Sir Ebenezer Howard, *Parliamentary shorthand writer and visionary.*
Impermeable millboard (commercial building material)
D M Sutherland, *former employee at a firm of Edinburgh paper mill engineers. In 1882 he had established a factory at Sunbury-on-Thames*, UK, *to produce millboard as a backing for Lincrusta-Walton. The millboard was made by hot-pressing waste paper.*
Krypton
William Ramsay, *Professor of Chemistry, University College, London, with* Morris William Travers.
Loudspeaker
Horace Short *of London. The compressed-air 'Auxetophone' was first used atop Blackpool Tower*, UK, *and the Eiffel Tower, Paris, to broadcast phonograph records of operatic arias.*
Motor car (enclosed body)
Renault's *2½hp 2-seater, France.*
Motor fire engine (self-propelled)
Cambie et Cie *of Lille, France, demonstrated at the French Heavy Autocar Trials held at Versailles.*
Neon
William Ramsay *and* Morris William Travers, *University College, London.*
Osmium Glühlampe (light bulb)
Karl Auer, Freiherr von Welsbach, *Austrian physicist and chemist.*
Parabellum (automatic pistol)
Georg Lüger, *Germany.*
Pedal lawn-mower
William Burnet, US.
Pneumatic tyre valve (one-piece replaceable core)
George H F Schrader, *New York,* USA.
Polonium
Pierre *and* Marie Curie, *France.*
Remote-radio-controlled boat
Nikola Tesla, *Yugoslav-born* US

*inventor. It was tested on the lake in
Central Park, New York City.*
Submarine (petrol-electric)
John Philip Holland US, *immigrant
Irish schoolteacher. His Mark VIII
craft was driven by a petrol engine on
the surface and an electric motor when
submerged, both installed on the same
shaft. It carried both a guncotton
torpedo tube and a submarine gun.*
Thermit process
Hans Goldschmidt, *industrial
chemist, Essen, Germany.*
Xenon
William Ramsay *and* Morris William
Travers, *University College, London.*

1899
Actinium
André Louis De Bierne, *chemist,
France.*
Artificial parthenogenesis
Jacques Loeb, *German-American
biologist.*
Aspirin (drug)
Dr Felix Hoffman *of Bayer* AG *of
Leverkusen, Germany, based on the
synthesization of acetylsalicyclic acid
by Karl Gerhardt, Alsatian chemist,
some 46 years before.*
Catalytic hydrogenation of oils
Paul Sabatier *and* Jean Baptiste
Senderens, *French chemists.*
Electric car (high speed)
Camille Jenatzy *of Belgium, who
achieved 66 mph on a road outside
Paris, beating his arch electric-car
rival, Count Gaston de Chasseloup-
Laubat of France.*
Electric-pneumatic signalling
Bishopsgate Station, London.
**Electric-wave wireless
telephone**
A Frederick Collins, US.
Foot-and-mouth disease vaccine
Friedrich August Johannes Löffler,
*Professor of Hygiene, University of
Griefswald, Germany.*
High-speed steel
Frederick Winslow Taylor *and*
Maunsel White, US.

**Inboard-outboard marine
engine drive**
Société du Propulseur Universel
Amovable *of Neuilly, France. It was
called the 'Lautonautile'.*
**Integral body-and-chassis
construction (motor car)**
Dr Frederick William Lanchester,
engineer, Edgbaston, Birmingham, UK.
Life-saving trunk
Herr Bunse, *Leipzig commercial
traveller.*
Oil-engined omnibus
London.
Pupin coil
Michael Idvorsky Pupin, *for
extending the range of long-distance
telephone calls. It was patented by the
American Telephone and Telegraph
Company in 1901.*
RDX (explosive)
Hans Henning, *German chemist;
fully developed 45 years later as
'Hexogen' or 'Cyclonite'.*
'Robin starch'
Reckitt & Sons Ltd, *Dansom Lane,
Hull,* UK.
**Tape-recorder (magnetisable
steel tape)**
Valdemar Poulsen, *Danish electrical
engineer, who demonstrated his
'Telegraphone' the following year at
the Paris Exhibition.*
Thoron isotope
Ernest Rutherford, *New Zealand-
born Professor of Physics, McGill
University, Montreal, Canada.*
Ticlocken coat
Thomas Burberry *for* UK *generals
during the Boer War.*

1900
Alkaline storage cells
Thomas Alva Edison, *Menlo Park,
New Jersey,* USA.
'Cineorama'
Raoul Grimoin-Sanson *at the Paris
Exposition. A battery of ten projectors
showed an elaborately hand-coloured
film on a completely circular screen*

1900

330 ft in circumference. The audience was seated in the middle.

Clementine (fruit)
Father Clément *of Oran, France.*

Cord tyre (motor cars)
The India Rubber, Gutta Percha and Telegraph Works Co. Ltd of Essex, UK. *Based on the invention of* John Fullerton Palmer *seven years before.*

Heavy-production grinding machine
the Norton Co., *Worcester, Massachusetts,* USA, *using artificial abrasives.*

Kala-azar parasite
William Boog Leishman, *assistant Professor of Pathology, Army Medical School, Netley, Hampshire,* UK.

Kodak 'Brownie' camera
5 shillings (25p) each.

Metal nuts and bolts (mass-produced)
Guest, Keen & Co.

Paper clip
Johann Vaaler, *Norwegian working in Germany.*

Photo-electric cell (practical)
Julius Elster *and* Hans Friedrich Geitel, *German physicists.*

Quantum Theory (general)
Max Karl Ernst Planck, *ordinary Professor of Theoretical Physics, Berlin, Germany.*

Radon (radiation emanation)
Friedrich Ernst Dorn, *chemist, Halle, Germany.*

Rigid airship (petrol-engined)
Lieutenant-General Count Ferdinand von Zeppelin, *Friedrichshafen, Germany. It was 420 ft long.*

Rolling pavement (variable speed)
Paris Universal Exposition. It transported 6.7 million visitors.

Submarine (petrol-electric)
Maxime Laubeuf, *France: the 'Narval'.*

Tryptophan (amino acid)
Frederick Gowland Hopkins, *biochemist, Prelector in Physiological*

Chemistry, Trinity College, Cambridge, UK, *and* S W Cole. *Six years later,* Hopkins *carried out his memorable experiment on animal nutrition.*

1901

Blood groups (four primary)
Karl Landsteiner, *Austrian scientist working in Switzerland, who classified them as A, B, AB and O.*

Diesel engine (reversible gear)
Frederic Dyckhoff *of Bar-le-Duc, France.*

Double-edged safety razor
King Camp Gillette, *former travelling hardware salesman of Fond du Lac, Wisconsin,* USA. *By 1906, with modifications by* William Nickerson, *Gillette's American Safety Razor Co. had sold some 90,000 razors and 12,400,000 blades.*

Gillette's safety razor

Electric typewriter
Dr Thaddeus Cahill *of Washington* DC. *Only 40 machines were built before the Cahill Writing Machine Co. ceased production.*

Fire engine (petrol-propelled)
Adler. *Water was discharged by gas pressure generated chemically.*

'First car radio'
Guglielmo Marconi, *Italian engineer working in London. A fully equipped Thornycroft steam wagon used for wireless telegraphy experiments.*

Grignard reagents
François Auguste Victor Grignard, *chemistry lecturer at Nancy, France.*

Hearing aid (electric)
Miller Reese Hutchinson *of New York. It was called the 'Acousticon'.*

Meccano
Frank Hornby, *Liverpool*, UK.

Motorcycle engine (centre-frame location)
Eugene *and* Michel Werner, *Russian émigrés working in Paris.*

Motor car mileometer
Bell Odometer by S H Davis Manufacturing Co., *Portland, Massachusetts*, USA.

Motor car speedometer
Thorpe & Salter Ltd, *Clerkenwell, London; calibrated from 0 to 35 mph.*

Multi-storey car park
City and Suburban Electric Carriage Co., *Piccadilly Circus, London – for owners of vehicles supplied by this company.*

Pre-selector gearbox
Dr Frederick William Lanchester, *Sparkbrook, Birmingham*, UK.

Vacuum cleaner
Hubert Cecil Booth, UK *civil engineer. His first machine was electrically powered and housed inside a horse-drawn carriage complete with 800-ft hose to go inside the house being spring-cleaned.*

Valine and proline (amino acids)
Emil Hermann Fischer, *German organic chemist in Berlin. He discovered oxyproline the following year.*

Wireless telegraphy (long-distance)
Guglielmo Marconi, *Italian engineer working in London. The letter 'S' was transmitted in Morse Code (...) from Poldhu, Cornwall, 2170 miles across the Atlantic to an aerial suspended from a kite in St John's, Newfoundland.*

1902
Air-conditioning
Willis H Carrier, US. *A 30-ton unit was installed in the Sackett-Wilhelms Lithographing & Printing Co's plant at Brooklyn, New York*, USA.

Air-cooled automobile
J Wilkinson *for* H H Franklin, US.

Alarm-clock tea-maker (spirit stove)
Frank Clarke, *gunsmith of Birmingham*, UK, *manufactured in 1904 by the Automatic Boiler Co., also of Birmingham.*

Disc brake (motor car)
Dr Frederick William Lanchester, *engineer, Sparkbrook, Birmingham,* UK.

Hormones
William Maddock Bayliss *and* Ernest H Starling, *his brother-in-law, physiologists at University College, London. They called their discovery 'secretin' but it was only the first of many hormones to be discovered in the human body.*

Indanthrene (non-fading dye)
James Morton, *textile manufacturer. He procured hundreds of samples of dyed fabrics and exposed them to light in a greenhouse, selecting those that did not fade.*

Intravenous anaesthetic
E Fischer, *biochemist, Germany. It was called 'Veronal'.*

Ivel agricultural motor
Dan Albone, *cycle manufacturer of Biggleswade, Bedfordshire*, UK, *after five years' development. Named after the local river, it remained in production until 1916.*

Kennelly-Heaviside layer (ionosphere)
discovered almost simultaneously by Arthur Edwin Kennelly, *Professor of Electrical Engineering, Harvard University*, USA, *and* Oliver Heaviside, *physicist working for the Great Northern Telegraphy Co., Newcastle-upon-Tyne*, UK.

Lawn-mower (petrol-engined)
James Edward Ransome, *Ipswich,* UK. *The 42-inch chain-driven machine*

1902

was powered by a 6 hp Simms petrol motor and equipped with a passenger seat.

Mercalli Scale of felt intensity (earthquakes)
Giuseppe Mercalli, *Italian geologist and seismologist.*

'Mischmetal' (cerium-iron alloy)
Karl Auer, Freiherr von Welsbach, *chemist, Austria.*

Motorboat (transatlantic)
New York Kerosene Co., USA. *Crewed by Captain William C Newman and son, 'Abiel Abbot Low' took 36 days to motor from New York to Falmouth, UK.*

Motor car (all-metal body)
the aluminium Napier 9, manufactured in London.

Motor caravan
Panhard *and* Levassor, *Paris, for Dr E E Lehwess, Germany, for an unsuccessful attempt to make the first motorised circumnavigation of the world.*

Motorcycle engine (V-twin, two-stroke)
Bichrone, *France.*

Overhead camshaft
Reginald W Maudslay *with* Alex Craig, *Coventry*, UK – *for both the first Maudslay car and commercial vehicle.*

Passenger excursion ship (steam-turbine)
'King Edward', launched on the Clyde.

Polygraph
James Mackenzie, *cardiologist, Burnley, Scotland; to measure arterial and jugular pulse simultaneously.*

Radium
Pierre *and* Marie Curie, *husband and wife, physicists at the School of Physics, La Sorbonne University, Paris, based on work by* Antoine Henri Becquerel *who had discovered 'Becquerel rays' in 1896. Having coined the term 'radioactivity', Marie died from its harmful side-effects.*

Solar engine (flat-plate/two-fluid system)
H E Willsie, *Olney, Illinois,* USA.

Straight-eight motor car engine (adapted)
CGV (Charron, Girardot and Voigt) *of Paris, using two of their four cylinder engines coupled together.*

Sulphur extraction process
Herman Frasch, *German-born chemist working for the Standard Oil Company at Louisiana,* USA.

Synchronous electric motor
Ernst Danielson, *Swedish engineer.*

Teddy bear
Simultaneously by Morris Mitchom, *Russian immigrant in a small sweet shop in Brooklyn, New York, with permission from* US *President Theodore ('Teddy') Roosevelt, and* Richard Steiff *at Giengen, Swabia.*

Track-circuit-controlled railway systems
London & South Western Railway Co., UK.

1903

Aeroplane (controllable wing-warping)
Orville *and* Wilbur Wright, *former bicycle engineers at Kill Devil Hill, Kitty Hawk, North Carolina,* USA. *Their first practical aeroplane was the 16 hp 'Flyer III' of 1905, which made 40 flights, some for over half an hour, at 35 mph.*

Barbiturate drugs
Emil Hermann Fischer *(Germany) and* Emil Adolf von Behring, *who was Professor of Hygiene at Marburg University, Germany.*

Box-kite boat
Colonel Samuel Franklin Cody, US, *who had a kite tow his boat across the English Channel.*

Cross-ply tyre
Christian Hamilton Gray *and* Thomas Sloper, *at the India Rubber, Gutta Percha and Telegraph Works Co., Ltd, Essex,* UK. *They also patented, but never marketed, a*

radial-ply tyre in 1913.
Instant photographic printing
G C Beidler, *and marketed by the* US
Rectigraph Co. in 1907.
Micro-cinematography
Martin Duncan *with* Charles Urban,
London, UK. *'The Unseen World' as
filmed by the Urban-Duncan Micro-
Bioscope was first shown at the
Alhambra Theatre, London.*
Motorboat (diesel-engined)
Rudolph Diesel, Frederick Dyckhoff
and Adrien Bochet. *'Petit Pierre' was
a canal burge belonging to Huchette
and Driout, French iron-founders.*
Motor-chemical fire engine
Messrs Merryweather *in close
collaboration with the* Tottenham
District Council, *London.*
**Motorcycle engine (in-line, air-
cooled, four-cylinder)**
FN armaments firm, *Belgium.*
'N-rays'
René Blondlot, *Professor of Physics at
Nancy University, France.*
Nitrogen-from-air process
Kristian Olaf Birkeland *and* Samuel
Eyde *of Kristiania, Norway.*
Postage-franking meter
Karl Uchermann, *Norwegian
inventor, manufactured by Krug
Maskinfabrik of Kristiania.*
Saloon car (enclosed)
Duryea Co. *of Coventry,* UK.
Seamless bust support
Kate Morgan *of London.*
**Skyscraper (reinforced
concrete)**
the Ingalls Building, Cincinnati, USA,
with sixteen storeys.
**Steam-turbine ship
(transatlantic)**
the small yacht, 'Emerald'.
Straight-six motor car engine
Jacobus Spyker, *De Industrieele
Maatschappij, Trompenburg,
Netherlands.*
**String galvanometer
(electrocardiograph)**
Willem Einthoven, *Professor of
Physiology at Leyden, Holland.*

Swinging axles
Edmund Rumpler, *engineer,
Germany.*
**Trimaran (steam-engined/
aerially propelled)**
Lawrence Hargrave *of Stanwell
Park, Australia.*
Ultramicroscope
Richard Adolf Zsigmondy, *Austrian
chemist.*
V-8 motor car engine
Clément Ader, *France.*

1904
**Air-cushion hydrofoil (swing-
wing)**
Clément Ader, *France. It worked well
but was too complicated and unwieldy
for practical use, so was presented by
its inventor to the Conservatoire des
Arts et Metiers, Paris.*
Caterpillar/crawler tractor
Benjamin Holt, US, *and* David
Roberts *with* Ruston Hornsby, UK.
Double-sided gramophone discs
International Talking Machine Co.,
*Weissensee, near Berlin – called
'Odeon' records.*
'Emanium X'
Friedrich Oskar Giesel, *chemist,
Germany.*
Forced lubrication (motor car)
Dr Frederick William Lanchester,
engineer, Birmingham, UK.
Heckelphone
Wilhelm Heckel, *musical instrument-
maker, Biebrich-am-Rhein, Germany.*
Ivel armoured car
Dan Albone *of Biggleswade,
Bedfordshire,* UK. *With steel armour-
plating, supplied by Cammel, Laird &
Co. of Birkenhead, to cover the three-
wheeled vehicle, it was a development
of the Ivel tractor.*
Kapok lifebelts
*made from a vegetable fibre found
around Java in the Dutch East Indies
that was impervious to water and
highly buoyant.*
Motor crash helmet
'casques', used during a motor race at

the Parc des Princes, France.
Motorcycle engine (parallel vertical twin)
Eugene *and* Michel Werner, *France.*
Novocaine (local anaesthetic).
Parabolograph
H Payne *of London University, with a grant from the Drapers' Company. It was designed and engineered by Swiss instrument-maker, Coradi.*
Pinacyanol (infra-red sensitising agent)
Homolka.
Rolling-road testbed dynamometer
Serge Berditschewsky Apostoloff, *Russian immigrant working in London.*
Rubber engine mountings
Dr Frederick William Lanchester, UK *engineer, Sparkbrook, Birmingham, UK.*
Ship (diesel-engined)
Messrs Nobel Brothers *at the Kolomna Works, St Petersburg, Russia. The 'Wandal', an oil tanker, was used on the Caspian Sea.*
Slow-motion film
Lucien Bull *and* Henri Nogues *at the Nancy Institute, France.*
Snow-chains (car tyres)
Harry D Weed *of Canastota, New York.*
Submarine (diesel-engined)
the French Navy's 'Aigrette' and 'Cicogne'.
Tantalum light filament (incandescent lighting)
W von Bolton, USA.
Telescopic motorcycle fork
simultaneously by Terrot *of France and* Alfred A Scott, *dyer's technician, Yorkshire, UK.*
Thermionic valve (vacuum tube)
John Ambrose Fleming, *Professor of Electrical Engineering at University College, London. This made possible long-wave radio transmissions. Fleming also invented the potentiometer.*

Fleming's thermionic valve

1905
Amphibian vehicle (petrol-engined)
Fournier, *France.*
Capillary motor-regulating mechanism (medical)
Schack August Steenberger Krogh, *marine biologist, University of Copenhagen, Denmark, following a study of the respiration of frogs.*
Dérailleur gear (bicycle)
Paul de Vivie, *France.*
Detachable wire wheel (motor car)
J V Pugh *of the Rudge-Whitworth Co., UK.*
Diesel engine (reversible two-stroke)
Sulzer Brothers, *Sweden.*
Electric motor horn
Miller Reese Hutchinson *of New York.*
Electric strip lighting
D McFarlan Moore *of Moore Electric Co., London – used for advertising.*
Electromagnetic seismograph
Prince Boris Borisovich Golitsyn, *physicist, Russia.*
Engine bearings (tapered oil film)
A G M Michell, *to avoid mechanical*

friction.
Fire-extinguisher (chemical foam)
Professor Alexander Laurent, *St Petersburg, Russia.*
Float-mounted glider (man-carrying)
Gabriel Voisin, *France. It took off from the River Seine, towed by 'La Rapière', a racing motorboat.*
Fluid coupling (hydraulic centrifugal clutch)
Hermann Fottinger; *developed by Vulcan of Hamburg for marine and diesel railway use.*
Gas turbine (workable)
René Armengaud *and* Charles Lemale, *Société des Turbomoteurs, Paris.*
Hydrofoil ladder (hull attachment)
Enrico Forlanini, *Italy. In 1906 Forlanini tested his hydrofoil boat on Lake Maggiore. It lifted clear of the water and reached a speed of 38 knots using a 75 hp engine.*
Hydroxy citronellal (synthetic perfume)
having the scent of lilies of the valley.
Juke-box (pre-selective)
John C Dunton *of Grand Rapids, Michigan, USA. It gave the listener a choice of 24 different Edison phonographic recordings.*
Liquid metal polish
W H Slack *for Reckitt & Sons Ltd, Hull, UK – 'Brasso'.*
Pneumatic bumpers (motor car)
Frederick R Simms *at his Kilburn factory, North London, for the 20 hp Simms-Welbeck car.*
Pneumatically operated semaphore indicator arms (motor car)
Fritz Berger *of Berlin. They were self-cancelling.*
Pressure cooker (aluminium)
Presto Co. *of Eau Claire, USA.*
Safety glass
John Crewe Wood, *solicitor of Swindon, Wiltshire, UK, who fitted it*

to his Peugeot Bébé.
Silicones
Frederic Stanley Kipping, *Professor of Chemistry, University College, Nottingham, UK.*
Special theory of relativity
Albert Einstein, *German-born, Swiss-Jewish mathematical physicist, whilst examiner at the Swiss Patent Office.*

1906

Alexanderson high frequency alternator:
Ernst Frederik Werner Alexanderson, *Swedish-born engineer, General Electric Co. Schenectady, New York, USA.*
Animated cartoon film
simultaneously by James Stuart Blackton *for the Vitagraph Co. of New York, USA, and* Walter Booth *for the Charles Urban Trading Co., London.*
Automatic flashing apparatus (lighthouses)
Nils Gustaf Dalén, *engineer-manager, Swedish Carbide and Acetylene Co. He also invented 'Solventil' and 'Agamassin'.*
Battleship (steam-turbine)
HMS *'Dreadnought'.*
Carbon suboxide
Otto Paul Hermann Diels, *Professor of Chemistry, Berlin University, Germany.*
'Cherry Blossom' (shoe polish)
Chiswick Soap & Polish Co., *London.*
Chromatography
Mikhail Tswett, *Russian botanist, Warsaw University.*
Constant-volume gas turbine
Dr Hans Holzwarth. *The first machine was built in Hanover, Germany, two years later.*
Crystal radio detectors (silicon, galena etc.)
H H C Dunwoody *and* G W Pickard, *US.*
Electric motor horn
United Motor Industries Ltd,

Sherbourne Works, Coventry, UK – called the Wagner Electric Motor Horn.

Electric washing-machine
Alva J Fisher, *Chicago, Illinois, USA. It was marketed in 1910 by the Hurley Machine Co. as the 'Thor'.*

Ellipsoidal-wing floatplane
Gabriel Voisin *and* Louis Blériot, *France. Trials on Lake Enghien, Belgium, were disastrous.*

Growth-stimulating vitamins
Frederick Gowland Hopkins, *biochemist, Prelector in Physiological Chemistry, Trinity College, Cambridge, UK.*

Heat theorem (third law of thermodynamics)
Hermann Walther Nernst, *Professor of Physical Chemistry, Berlin.*

Hookless zip-fastener (matching metal locks on a flexible backing)
Dr Gideon Sundback, *Swedish émigré electrical engineer, working for the Judson Company, Chicago. In 1913 he further developed a zip-fastener manufacturing machine.*

Humidity control and dust filter (air conditioning)
Stuart Warren Cramer, *mill-engineer, Charlotte, North Carolina, USA.*

Ionium
Bertram Borden Boltwood, *Professor of Physics, Yale University, USA.*

Kinemacolor (motion pictures)
George Albert Smith *of Brighton, UK, sponsored by businessman Charles Urban. Regular public screenings were held three years later at Urbanora House, Wardour Street, London.*

Lutetium
Georges Urbain, *assistant Professor of Analytical Chemistry, la Sorbonne. Latin 'Lutetia' is the ancient name of Paris, where Urbain was working.*

Marine outboard engine (mass-produced)
Cameron B Waterman, George Thrall *and* Oliver E Barthel, *Detroit,*

Michigan, USA.

Nichrome
Albert Marsh *of Lake County, Illinois, USA. This was the ideal alloy for electric-fire elements.*

Permanent waving (hairdressing)
Karl Ludwig Nessler, *German-born hairdresser working in London.*

Pneumatic-tyred undercarriage (aircraft)
Trajan Vuia, *Italian, with his tractor-propeller monoplane, the Vuia No. 1.*

Public radio broadcast
Reginald Aubrey Fessenden, *Canadian-born professor at his Brant Rock Station, Massachusetts, USA.*

Röntgen radiation of the elements
Charles Glover Barkla, *research physicist and chorister, University of Liverpool, UK.*

Sound-on-film motion pictures
Eugen Augustin Lauste, *French engineer working in Brixton, London. Four years later he produced a workable system using an electromagnetic recorder and string galvanometer.*

SOS (distress signal)
British Marconi Society *and* German Telefunk *organisation at the Berlin Conference. Formally introduced in 1908.*

Steamcar (high-speed)
F O *and* F L Stanley, *whose vehicle, driven by Fred Marriott, reached 127.5 mph on Ormonde Beach, Daytona, Florida, USA.*

Telephone amplifier valve
Robert von Lieben, *Vienna, Austria.*

Triode amplification tube
Professor Lee de Forest *of New York, USA. It was called the 'Audion'.*

Wheeled undercarriage (aircraft)
Alberto Santos-Dumont, *Brazilian engineer-aviator working in Paris, for his '14-Bis'.*

1907

Banked motor-racing circuit (custom-built)
Messrs Donaldson *(railway technician)*, Holden *(colonel, Royal Engineers)* and L G Mouchal, *as commissioned by Hugh Fortescue Locke-King of Brooklands Estate, Weybridge, Surrey,* UK.

Cinema projection system
The *'Kinedrome'* – D J Bell *and* A S Howell, USA.

Detergent (household)
Henkel et Cie, *Düsseldorf. Based on the dry soap powder process of* Professor Hermann Geissler *and* Dr Hermann Bauer, *it was called 'Persil'.*

Helicopter (petrol-engined)
independently by Frenchmen Paul Cornu, *with a twin-rotor design, and* Louis Breguet, *with four rotors on outriggers. Cornu and his passenger hovered at 5 ft for over a minute.*

Hydroplane (aerially propelled)
Comte de Lambert *of France and* Horatio Phillips, UK. *Powered by a Serpollet steam engine with a flash boiler, it was tested on the River Seine, France.*

Monoplane
Louis Blériot, *France, whose 50 hp Antoinette-engined Type VII made a long enough flight to be deemed airworthy. Blériot's 1909 cross-Channel flight underlined the monoplane's stability. Blériot also invented the aileron.*

Multiplane (venetian-blind type)
Horatio Phillips, UK. *This, his second attempt at a multiplane machine, had 50 horizontal winglets and was powered by a 22 hp engine. It flew to 500 ft during a flight test over Norbury,* SE *London – according to Phillips.*

Ocean liners (steam-turbine engines)
Cunard Shipping Co.'s *'Lusitania' and 'Mauretania'.*

Spangler's upright vacuum cleaner, trade-named 'Hoover'

Rotary aero engine (production model)
Laurent Seguin, *France. It was marketed as the 50 hp 'Gnome'.*

Supercharged motor car
Lee S Chadwick and his assistant *at Pottstown, Pennsylvania,* USA – *'The Great Chadwick Six'.*

Upright vacuum cleaner (dustbag attached)
J Murray Spangler, *janitor in a department store at New Berlin, Ohio. Spangler's patent rights were purchased by former harness-maker,* W H Hoover, *who produced a commercial model the following year.*

1908

Automatic typewriter
T A McCall *of Ohio,* USA. *It was controlled by a pianola-type roll.*

Bakelite
Dr Leo Hendrik Baekeland, *Belgian-born inventor working in Yonkers, New York. At temperatures of over 100°C, Baekeland experimented with formaldehyde and phenolic bodies to produce a synthetic shellac from simple molecules. After developing his process into a three-stage reaction, he took out patents and in 1912 became President of the Bakelite Coporation with the slogan 'the material of a thousand uses'.*

BCG (tuberculosis) vaccine
Leon Charles Albert Calmette, *Director of the Pasteur Institute, Lille, France, with* Dr Camille Guérin.

Cellophane
Dr Jacques Edwin Brandenberger, *Swiss-born French dye-chemist at Thaon-les-Vosges – a transparent wrapping material. The Comptoir de Textiles Artificiels, then the largest rayon producer, agreed to finance him. A company called La Cellophane was formed and machine production began in 1912.*

Crankshaft torsion dampers
Dr Frederick William Lanchester, UK *engineer, Sparkbrook, Birmingham,* UK.

Diesel engine (reversible four-stroke)
Nobel Brothers, *oil distributors, St Petersburg, Russia.*

Filter coffee machine
Frau Melitta Bentz, *Dresden, Germany.*

Gyroscopic compass
Dr Hermann Anschütz-Kämpfe *and* Dr Max Schuler, *Kiel, Germany, for polar exploration. The Sperry-Ford version followed in 1910.*

Hydroplanes (multi-step/petrol-engined)
William Henry Fauber, *former bicycle engineer of Chicago, Illinois,* USA.

Language-teaching system
Thódore Rosset *of Grenoble, France. Using gramophone discs and ciné films to teach French to foreigners.*

Liquid helium
Heike Kamerlingh-Onnes, *Dutch physicist at the Cryogenic Laboratory, Leyden.*

Nickel-cadmium storage cells
Jungner, *Sweden.*

Nickel-iron alkaline storage battery
Thomas Alva Edison, *Menlo Park, New Jersey,* USA.

'Noiseless' typewriter
W P Kidder *and* C C Colby *of Quebec, Canada.*

Oil (south-west Persia)
William Knox d'Arcy, *wealthy Englishman, who had made a fortune in gold-mining in Australia. The Anglo-Persian Oil Company was formed in 1909, following the strike at Masjid-i-Sulaiman.*

Paper cups (for drinking)
Public Cup Vendor Co., *New York.*

Poliomyelitis
Karl Landsteiner, *pathologist, Austria.*

Pre-transfusion blood tests
Dr Reuben Ottenberg, *New York.*

Safety film (cellulose acetate base)
Kodak, *Rochester, New York,* USA.

Sedimentation equilibrium
Jean Baptiste Perrin, *Professor of Physical Chemistry, Paris University.*

Ship-steadying gyroscope
Dr E O Schlick *and* Pierre Schlowsky.

Silencer (firearm)
Hiram Stevens Maxim, *London.*

Solar thermoelectric device
W Zerassky, *Russian engineer.*

Thought machine (needle-deflecting-scale)
Edmonde Savary d'Odiardi, *Paris.*

Tungsten filament (ductile)
William D Coolidge, *General Electric Co., Schenectady,* USA.

V-Eight motorcycle
Glen L Curtiss, US.

1909

Barany indication test
Robert Barany, *Austro-Hungarian lecturer in Otology, the Ear Clinic, Vienna University.*

Doppler effect in canal rays
Johannes Stark, *physicist at the Technische Hochschule, Aachen, Germany.*

Double-decker bus (enclosed)
Widnes Corporation, UK.

IUD (intra-uterine contraceptive)
Dr R Richter, *German physician. He devised a ring-shaped* IUD *out of silkworm gut.*

Salvarsan (compound 606)
Dr Paul Ehrlich *and* Hideyo
Noguchi, *bacteriologists, Frankfurt-*
am-Main, Germany. Nicknamed 'the
magic bullet', this arsenic compound
was the first chemical cure for disease
in this case, early syphilis. It was
manufactured by Hoechst of Germany.

Synthetic ammonia process
Fritz Haber, *Professor of Physical*
Chemistry, Karlsruhe, Germany;
industrially exploited in 1913 by Carl
Bosch at the Badische Anilin und
Sodafabrik's pioneer factory.

Typhus fever body louse
Charles Jules Henri Nicolle, *Director*
of the Pasteur Institute, Tunis.

1910

Aero-engined car
Fiat *of Turin, Italy. A 300 hp airship*
engine was installed in a motor car,
giving it a speed of 132.37 mph at
Ostende, Belgium.

Duralmin (metal)
Durener Metallwerke AG, *Germany.*

Gene theory
Thomas Hunt Morgan, *Professor of*
Experimental Zoology, Columbia
University, New York, during his
study of thousands of generations of the
fruit-fly Drosophila.

Hydro-aeroplane (seaplane)
Henri Fabre *of Marseilles, France. In*
his first flight from Martigues, Fabre
was airborne at 6 ft for 500 yards.

Neon lighting
Georges Claude, *French physicist. It*
was first used to illuminate the
peristyle of the Grand Palais during
the Paris Motor Show. Within four
years, some 150 neon signs had been
installed on buildings in Paris alone.

Powers' accounting machine
James Powers, US – *automatic card*
punch, sorter and tabulator.

Radio-direction finder
Ettore Bellini *and* Alessandro Tosi,
Italy; acquired by the Marconi Co.,
UK, *and first installed at Boulogne,*
France.

Spring-operated mouse-trap
James Henry Atkinson, *Leeds,* UK.

Tanker ship (diesel-engined)
'Vulcanus', built by the Netherlands
Shipbuilding Co. of Amsterdam for
operation by the Anglo Saxon
Petroleum Co.

Triplex laminated glass
Edouard Benedictus, *Paris.*

Tumour-inducing virus
Francis Peyton Rous, *associate of the*
Rockefeller Institute for Medical
Research, New York, USA.

1911

Aircraft carrier (adapted)
the US *Navy's cruiser, 'Pennsylvania',*
with a platform built over her stern.
The plane was landed by Eugene Ely.

Air-raid
Lieutenant Giulio Gavotti, *Italy, in*
an Etrich monoplane, dropped several
4½ lb Citelli bombs on both the Turkish
position at Ain Zara and the oasis at
Tagiura.

Airship-tethering mast
(offshore)
H B Pratt *of Vickers-Armstrong Ltd*
for the Cavendish Dock, Barrow-in-
Furness, UK, *to secure the 'Mayfly'*
airship.

Amphibious biplane (canard
configuration)
Gabriel Voisin. *Piloted by Maurice*
Colliex, it took off from Issy
aerodrome, France, landed on the
River Seine and returned to Issy. A
dozen models were acquired by both the
French and Russian Navies soon after.

Binet intelligence scales
Alfred Binet, *Director of*
Physiological Psychology at the
Sorbonne, France, with Theodor
Simon.

Blood groups A1 and A2
Karl Landsteiner, *Professor of*
Pathological Anatomy, Vienna.

Calculating machine (full
automatic multiplication and
division)
Jay R Munroe, *New Jersey,* USA.

1911

Cloud chamber
Charles Thomson Rees Wilson, *Scottish physicist at the Cavendish Laboratory, Cambridge*, UK.

Cosmic radiation
Victor Franz Hess, *Assistant at the Institute of Radium Research, Vienna Academy of Sciences*.

Diesel ship (transatlantic)
'Toiler', fitted with a Swedish engine.

Electric frying pan.

Harmonic balancers
Dr Frederick William Lanchester, *engineer, Sparkbrook, Birmingham*, UK.

Hydrofoil biplane
A Guidoni, *Italy, using a modified Farman F.1.*

Intertype (printing process)
Hermann Ridder.

Isotopes
Sir Joseph John Thomson *at the Cavendish Laboratory, Cambridge*, UK.

Machine-gun (air-cooled)
Isaac Newton Lewis, US *artillery officer. It was the first machine-gun to be fired from an aircraft – in 1912.*

Molecular beam
Louis Dunoyer de Segonzac *at Marie Curie's Laboratory, La Sorbonne, France*.

Motion-picture camera (electric drive)
E F Moy *of London. With a gyroscope to stabilise it and a displacement film-magazine, it replaced hand-cranked machines*.

'Nivea' cream
Beiersdorf GmbH, *Hamburg, Germany*.

Oil (underwater drilling)
Gulf Refining Co., *in Caddo Lake on the Louisiana-Texas border*, US.

Powerboat (aero-engined)
DeKorwin hydroplane, 'Soulier-Volant', powered by a US *Wright aero-engine. It raced at Monaco.*

Retractable undercarriage
Wiencziers *monoplane, Germany*.

Self-starter (automobile)
Charles F Kettering *at Dayton, Ohio*, USA. *The first 4,000 'Delco' starters were sold to Cadillac Motors.*

Theory of atomic structure
Ernest Rutherford, *New Zealander, and his Danish assistant*, Niels Henrik Bohr, *at the University of Manchester*, UK.

'US Patent No. 1,000,000'
Francis H Holton, *Akron, Ohio – described as 'vehicle tyre improvements'.*

1912

Activated sludge process (sewage)
Arden *and* Lockett, *Manchester*, UK.

Aerograph (pneumatic-drive cine camera)
Arthur Samuel Newman, *inventor, Highgate, London, for use on horseback by Count Proszynski of Russia.*

Cabin biplane
Igor Sikorsky, *Russia, aged 23. With four 100 hp engines, a 92 ft wingspan, an enclosed cockpit for pilot and co-pilot with dual controls, a luxuriously decorated cabin for sixteen passengers and even a toilet, this was the prototype of the modern airliner, even if it only made a few flights. A cabin biplane was also built by* Alliot Verdon Roe, UK, *during this year.*

Diffraction of X-rays by crystals
Max Theodor Felix von Laue, *Professor of Theoretical Physics at Munich, with his assistants*, Friedrich *and* Knipping.

Electric fire (fire-clay frame)
Charles Reginald Belling, *Enfield, Middlesex*, UK. *It was called the 'Standard'.*

Hydraulic hoist (device for dumping trucks)
Garfield Arthur Wood, *mechanic and boatbuilder, St Paul, Minnesota*, USA.

Liquid silver polish
Reckitt & Sons Ltd, *Hull*, UK – *'Silvo'.*

Locomotive (diesel-engined)
North British Locomotive Co. –
*1000 hp diesel-mechanical unit with a
direct drive Klose-Sulzer engine.*

Monocoque monoplane
Louis Becherau, *32-year-old French
engineer working for Armand
Deperdussin, wealthy silk broker. It
was called the 'Deperdussin'.*

Motor-narrowboat (custom-built)
Fellows, Morton & Clayton *of
Birmingham,* UK.

**Regenerative, oscillating
features of the triode**
simultaneously by Lee de Forest, US;
and Edwin Howard Armstrong, US;
and Fritz Löwenstein, US, Siegmund
Strauss, *Austria, and* A Meissner,
Germany.

Self-service grocery store
*Alpha Beta Food Market at Pomona,
and also Ward's Grocetaria in Ocean
Park, California,* USA.

**Stainless steel (chromium
alloys)**
simultaneously by Harry Brearley,
self-taught metallurgist, UK; *and*
Elwood Haynes, *automobile inventor
of Kokomo, Indiana,* US; *and* Edward
Maurer *and* Benno Strauss *of
Krupps, Germany.*

Synthesised butadiene
Chaim Weizmann, *German-born
chemist at Manchester University,* UK
– bacterial fermentation culture.

Synthetic wood glue
Dr Leo Hendrik Baekeland, *Belgian-
born chemist, General Bakelite
Corporation,* USA.

Vest pocket Kodak camera
USA; *almost 2 million sold by 1926.*

Vitamin B1 (thiamine)
Casimir Funk, *Polish-born biochemist
working at the Lister Institute,
London.*

1913
Autopilot
Lawrence Sperry, US, *on a Curtiss
floatplane.*

Bicycle (semi-bodyshelled)
Etienne Buneau-Varilla, *France. It
was called the 'Vélo Torpille' (torpedo
bicycle).*

Crossword puzzle
Arthur Wynne, *Liverpool-born. It
was first published in the weekend
supplement of the 'New York World'.*

Diesel-electric railcar
*Mellersta Sodermanlands Railway,
Sweden.*

**Dumet (iron-nickel alloy wire
for incandescent lamps)**
Dr Colin Garfield Fink,
*electrochemist, Head of the Research
Laboratory, Chile Exploration Co.,
New York City,* USA.

**Electric self-starter
(motorcycle)**
Indian Co., US. *Called the
'Dynastart', it was used on their
Hendee Special.*

Gas-filled tungsten lamp
Irving Langmuir, *Research
Laboratory, General Electric Co.,
Schenectady, New York,* US.

Hydrocycle (aerial propeller)
Charolais-Favelier, *France.*

Isotopic labelling technique
Georg von Hevesy (*Hungary*) *and*
Friedrich Adolf Paneth (*Austria*)
whilst staying in Vienna.

**Moving assembly line for
automobile magnetoes**
Henry Ford *of Ford Motor Company,
Detroit,* USA, *for the mass-production
of his Model T utility automobile, 'Tin
Lizzie', alias 'Flivver'. By 1914 the
system had been extended to the
assembly of the entire car. By 1927
over 15 million 'Tin Lizzies' had come
off the production line and been sold.*

Protactinium
Frederick Soddy, *Lecturer in
Physical Chemistry at Aberdeen, with*
John A Cranston, UK.

Quantum theory (sub-atomic)
Niels Bohr, *Danish physicist,
following work with* Thomson *at
Cambridge and* Rutherford *at
Manchester – applying the theory to*

1914

sub-atomic physics.

Synthetic detergent
A Reychler, *Belgian chemist. It was marketed in 1917 as 'Nekal'.*

Vitamin A
simultaneously by Thomas B Osborne, *biochemist,* US, *with* Lafayette B Mendel, *Professor of Physiological Chemistry, Yale University, and* Elmer V McCollum *with* M Davis. *at the Wisconsin Agricultural Experiment Station,* USA.

Wireless communication to an aeroplane
Received by Elmo Pickerill in a Curtis biplane, up to a height of about 1 mile.

Woolworth Building (60 storeys)
Cass Gilbert, *American architect. Broadway, New York City,* USA.

1914

Aluminium-foil bottle caps
Josef Jonssen *of Linköping, Sweden.*

Brassière
Miss Mary Phelps Jacob *(later glamorous heiress,* Mrs Caresse Crosby*), whilst a débutante in New York,* USA. *Miss Jacob's patent for the 'backless brassière' was eventually acquired by the Warner Brothers Corset Company for a mere $15,000.*

Curved windscreens (automobile)
Kissel, USA.

Dye-coupler colour process
Dr Rudolph Fischer *and* Siegrist, *Berlin, Germany – one of the factors which ultimately made colour photography a practical reality. The weakness in the Fischer-Siegrist process was the lack of diffusion control between the layers of sensitising dyes.*

Electric traffic light (red)
Alfred A Benesch, *Cleveland, Ohio,* USA, *who formed the Traffic Signal Co.*

'Fléchettes' (steel darts)
jettisoned by Royal Flying Corps pilots from their aircraft over enemy lines.

Giant mortar (gun)
'Big Bertha', built by the Skoda Works in Austria-Hungary and used by the Germans in battering the forts at Liège and Namur, Belgium. Capable of firing a 1980 lb shell, it was nicknamed after Frau Bertha von Bohlen, head of the Krupp armaments factory.

Grand Prix engine (desmodromic valve gear)
Automobiles Delage, *Courbevoie, France.*

Insulating board
Karl Munch, *Germany.*

Leica (35 mm film still camera)
Oskar Barnack, *microscope designer at Leitz in Wetzlar, Germany. It went into production ten years later.*

Oskar Barnack's original 35mm Leica camera

Multi-engined heavy bomber
Igor Sikorsky's *'Ilya Mourometz', for the Imperial Russian Air Corps.*

Panama Canal
John F Stevens, US *engineer, and* Lt Colonel George W Goethals, *in charge of the* US *Army engineering detachment.*

Passenger-carrying aeroplane (commuter)
Benoist Flying Boat, carrying passengers between St Petersburg and Tampa, both in Florida, USA.

Tear gas (xylyl-bromide)
Dr von Tappen, *Berlin. It was used in 1915 by the Germans against the Russians at Bolimow, Poland.*

1915

Airship air-raid
'Zeppelin L-3,' over Great Yarmouth, UK. *It killed two people.*

All-metal aeroplane (cantilever wing)
Hugo Junkers, Dr Mader *and* Otto Reuter *at Dessau, Germany. The Junkers J.1 was nicknamed the 'Tin Donkey'.*

Amplitude modulation (AM)
Hendrick Johannes van der Bijl, *South African physicist, and* Raymond A Heising, *electrical engineer,* US, *at the Western Electric Co., New York.*

Bacteriophage
Frederick William Twort, *Professor of Bacteriology, University of London.*

Carrel-Dakin treatment
Alexis Carrel, *French Army Surgeon, and* Henry Drysdale Dakin, UK *chemist, as a solution for treating wounds received in combat.*

Chlorine gas (weapon)
suggested by Sergeant Fritz Haver, *Jewish chemist. It was used by Germany against the Allied Forces at Ypres, West Belgium, so was nicknamed 'Ypérite' by the French troops, who tried tying pads of cotton dipped in a chemical solution over nose and mouth to avoid harmful effects.*

Gas mask (veil mask respirator)
Robert Davis, Siebe Gorman & Co., *Lambeth, London.*

Equal-distance bomb sight (for aircraft)
Warrant Officer F W Scarff, *Royal Navy, and soon adopted by the Royal Naval Air Service,* UK.

Motor scooter
Auto-Ped Co. *of New York,* USA.

Periscope rifle
Lance-Corporal Beech *from Sydney, Australia, whilst fighting at Gallipoli, Turkey.*

Propeller-synchronised machine-gun
Anthony Herman Gerard Fokker, *Dutch engineer working for the Germans. This enabled the pilot to fire his machine-gun without destroying the wooden blades of his propeller. The scourge of the Fokker Eindecker*

monoplane with its synchronised gun lasted nine months before a comparable device was produced by George Constantinesco for the Allied Forces.

Radiotelephone (long-distance)
Western Electric Telephone and Telegraph Co., USA, *with long-distance calls from Arlington, Virginia to San Francisco and Hawaii, then three weeks later across the Atlantic to Paris.*

Tank (tracked, armoured vehicle)
RNAS Lieutenant Walter Gordon Wilson, *Armoured Car Squadron, working with* William Tritton *of Fosters in Lincoln,* UK. *Following their prototype, 'Little Willie', the lozenge-shaped boiler-plated 'Mother' was so successful that 100 ironclads were ordered. For secrecy's sake, they were called 'Water Carriers', hence the word 'tank'. By 1917 some 450 improved British Mark IV tanks attacked the Hindenburg Line at Cambrai, capturing 7500 prisoners and 120 guns. Each tank had an eight-man crew.*

'Tank' – tracked armoured vehicle

Tank (French design)
Jean-Baptiste Estienne, *French artillery officer.*

Thermostatic oven regulator (gas stoves).

1916
Concrete shell structure
Eugène Freyssinet, *French architect.*

Albert Einstein

Two giant airship hangars at Orly Airport, France, were completed by 1924.

General theory of relativity
Albert Einstein, *Director of the Kaiser Wilhelm Physical Institute in Berlin.*

Long-wave M-lines (X-ray spectroscopy)
Karl Manne Georg Siegbahn, *Deputy Professor of Physics at Lund University, Sweden.*

Mechanical windscreen wipers (automobile)
Willys-Knight, *New York City*, US.

Microphone (condenser type)
E C Wente.

Nissen hut
Peter Norman Nissen, *Canadian-born officer commanding 29 Company Royal Engineers. A prototype was erected at Hesdin, France.*

Open-air public address system
Bell Telephone Co., *Staten Island, New York*, USA.

Organic micro-analysis
Fritz Pregl, *Professor of Applied*

Medical Chemistry, Innsbruck, Austria.

Radio tuning device (selective)
Ernst Frederik Werner Alexanderson, *General Electric Co., Schenectady, New York*, USA.

Vibraphone (musical instrument)
Hermann Winterhoff, *of the Leedy Drum Company*, US.

1917

Chernikeef ship's log
Captain B Chernikeef *of the Russian Naval Hydrographic Department.*

Echo-sounder
Paul Langevin, *French physicist, and* Professor Robert W Boyle. *Called* ASDIC *(Anti-submarine Detection Investigation Committee) it was ready for use by the Royal Navy in 1918.* SONAR, *the* US *version, was developed concurrently.*

General treatment for paralysis
Julius Wagner von Jauregg, *Austrian neurologist and psychiatrist.*

Machine-gun (water-cooled/ recoil-operated)
John Moses Browning, *inventor, Ogden, Utah*, USA.

Mustard gas
German scientists. *In 1918, during ten days of the Somme offensive, the Germans used some 500,000 mustard gas shells. The newly developed, unwieldy gas masks used by the Allied Forces were of little help.*

Roll-towel cabinet
George Steiner *of the American Linen Supply Co., Salt Lake City, Utah,* USA.

Superheterodyne radio receiver
Lucien Lévy, *France.*

Tractor (mass-produced)
Ford Motor Co., *Detroit*, USA. *By 1926, 70,000 'Fordsons' were coming off the assembly line per year, 75 per cent of all tractors made in the* USA.

Very High Frequency (VHF)
Guglielmo Marconi, *Italian electrical engineer, experimenting near*

Caernarvon, Wales, UK.
Windscreen wipers (electric)
Ormond Edgar Wall, *Hawaiian
dentist.*

1918
Anti-tank gun
German armaments engineers.
Domestic refrigerator (electric-power)
Nathaniel Wales *and* E J Copeland,
US. *It was called the 'Kelvinator'.*
Food mixer (electric)
Universal Co., US.
Non-rigid airship (helium-filled)
US Navy.
Practical electric clock (AC)
Henry Ellis Warren, *electrical
engineer, Ashland, Massachusetts,*
USA.
Sheet glass
Emile Fourcault *in Belgium and*
Irving Colburn *in the* USA.
Traffic lights (red-green-amber)
New York City, USA. *They were*

The 'Universal' electric food-mixer

*electrically powered but worked
manually from a 'crow's nest'.*
Ventriculography
Walter Edward Dandy, *Professor of
Neurological Surgery, Johns Hopkins
Hospital,* USA.
Vitamin D (antirachitic factor)
Edward Mellanby, *Professor of
Physiology, University of London,* UK.

1919
Aircraft (transatlantic)
Vickers Vimy, piloted by Captain
John Alcock *and* Lieutenant Arthur
Whitten-Brown *from Newfoundland
to Ireland, taking almost sixteen hours.*
All-welded ship (offshore)
Cammell Laird *of Birkenhead,* UK,
*whose 'Fullager' was built for the
Anchor Brocklebank Line of
Liverpool.*
Atomic nuclei bombardment
Sir Ernest Rutherford, *Cavendish
Professor of Physics at Cambridge
University,* UK.
Compound epicyclic gear train
Walter Gordon Wilson, UK. *First
tried experimentally with a 1927
Vauxhall, it was later used in his pre-selector gearbox.*
**Electrets (permanent
electrostatic fields)**
Mototaro Eguchi, *Japan.*
Flying boat (transatlantic)
Glenn Curtiss's *NC-4 from Long
Island,* USA *to Lisbon, Portugal.*
Grease-filled 'nipple'-gun
Edward Coe Critchlow, *production
superintendent of the Union Oil Co. of
California,* USA.
Hydrofoil (high-speed)
Alexander Graham Bell. *His 60 ft
'Hydrodome IV' achieved a world
water speed record of almost 71 mph.*
Mass spectrograph (double-focusing)
Francis William Aston, *scientist at
the Cavendish Laboratory,
Cambridge,* UK, *to investigate isotopic
structures of elements.*

Microfilm-reading machine
Rear-Admiral Bradley A Fiske, US
Navy, New York. It was called the
'Fiskeoscope'.
Neutrodyne receiver
Alan Hazeltine, *Professor of*
Electrical Engineering at Stevens
Institute, Hoboken, New Jersey, USA.
Parachute (manually operated)
Leslie 'Ski-Hi' Irvin *and* Floyd
Smith, *former* US *circus acrobats and*
balloonists. The first successful jump
was from a de Havilland 9 biplane
from a height of 1500 ft. The Irvin Air
Chute Co. was formed in Buffalo soon
after.
Servo-assisted, four-wheel
brakes
innovated for the Type H6 Hispano-
Suiza automobile.
Straight-eight automobile
engine
Fred Duesenberg, *Indianapolis,* USA.
Thyroxine
Edward Calvin Kendall, US *chemist.*
It was synthesised seven years later by
Charles Robert Harington, UK
chemist.
'Trigger relay' (electronic valve
as switch)
William Henry Eccles *and* Richard
Jordan, *physicists at the Imperial*
Wireless Telegraphy Commission, UK.

1920
Airscrew-propelled motor car
Leyat, *France.*
Autopilot equipment
the Aveline-Stabiliser, as used on the
Handley-Page 0/10, built and flown
from Cricklewood, North London.
Catgut (surgical thread).
Electrical recording process
Lionel Guest, *stockbroker and* H O
Merriman *of Paddington,London.*
Fire-fighting engine (foam-
equipped)
Foamite Childs Corp., US.
Light bulb and vacuum tube
mass-production machine
Corning Glass Works, Illinois, USA.

Capable of making 20-30,000 glass
bulbs per day.
Microphone (double-button type)
Bell Telephone Laboratories, Holmdel,
New Jersey, USA.
Radio station (regular
broadcasts)
Westinghouse's KDKA, *Pittsburgh,*
Pennsylvania, USA.
Rotary hoe (steam-powered)
Cliff Howard, *farm-engineer,*
Gilgrandra, New South Wales,
Australia. It was later called the
'Rotovator'.
Tea-bags
Joseph Krieger, *San Francisco,* USA.
X-ray analysis of crystal
structure
William Henry Bragg, *Quain*
Professor of Physics, University
College, London, and his son, William
Lawrence Bragg *at Manchester*
University – using their spectrometer
specially developed for this purpose.
Windscreen wipers (vacuum-
operated)
William Mitchell Folberth, *auto-*
mechanic, Cleveland, Ohio, USA*;*
manufactured by the Trico Corp. of
Buffalo.

1921
Aerial crop-dusting
Lieutenant John B Macready,
piloting a Curtiss JN6 light aircraft
over a catalpa-infested grove in Troy,
Ohio, USA. *It had been suggested by C*
R Neillie, Cleveland entomologist.
Aerocycle
Edouard Nieuport, *former bicycle*
racer and aircraft manufacturer,
France. Pedalled by Gabriel Poulain,
the Aviette flew 10.54 metres at a
height of 1.5 metres above the
Longchamps racecourse, so winning the
Peugeot Prize.
Automobile reversing light
Wills-Sainte Claire Co., *Marysville,*
Michigan, USA.
Glutathione
Frederick Gowland Hopkins,

Professor of Biochemistry, Cambridge, UK.

Hydraulic four-wheel brakes
Duesenberg Motor Co., Indianapolis, USA.

Insulin (animal)
Frederick G Banting and Charles H Best at J J R Macleod's laboratory, Toronto University, Canada.

Ion-radiated food (practical use)
Schwarz, US, to kill 'trichinella spiralis' bacillus in meat.

Lie detector
John Larsen, medical student, California, USA

Ley lines
Alfred Watkins, mill owner and photographer, Herefordshire, UK, who claimed that the pre-Roman trackways of Britain were constructed in straight lines, marked out by mystical and religious temples and monoliths. Following publication of his 'Ley Hunter's Manual' in 1927, the Straight Track Postal Portfolio Club was formed to re-plot the network of 'leys'.

Magnetron
Dr Albert Hall, General Electric Co., Schenectady, New York.

'Neracar'
J Neracher, Syracuse, New York, USA.

'Robot'
the word was used by Karel Capek, Czech playwright, in his play 'RUR' (Rossum's Universal Robots) – from the Czech 'robota' (work).

Slotted aircraft-wing (anti-stall-and-spin device)
simultaneously by Gustav Viktor Lachmann, engineer at Professor Ludwig Prandtl's Aerodynamic Institute, Göttingen, Germany and Frederick Handley Page, aircraft manufacturer, Cricklewood, London.

Tetraethyl lead (anti-knock petrol compound)
Charles Franklin Kettering and Thomas Midgley, General Motors Research Laboratory, USA. A practical manufacturing process for 'premium' gasoline was developed by Doctors Charles A Kraus and Conrad C Callis for the Standard Oil Company.

Triple-hydro-triplane (giant flying-boat)
Gianni Caproni of Italy. With eight aero-engines, 'Capronissimo' only made one successful test flight, of less than 2 km.

Wirephoto
Western Union Cables, USA.

1922
Aircraft carrier (custom-built)
the Japanese Navy's 'Hosho', followed nine months later by HMS 'Hermes'.

Automatic doors (underground railway carriages)
Cammell Laird of Birkenhead, UK – for London's Piccadilly Line.

Clavilux
Thomas Wilfrid, Danish-born singer and scientist working in the USA. Rather than make sound, the Clavilux had its own colour orchestration with works composed and performed by Wilfrid such as his Opus 39 Triangular Etude. In 1926 Leopold Stokowski conducted the Philadelphia Orchestra playing Rimsky Korsakov's 'Scheherezade' to a synchronous performance by Wilfrid on his 'colour-organ'.

Continuous hot strip rolling (steel)
John B Tytus, working for the American Rolling Mill Co., with assistance from Charles Hook, chairman of that company. By 1923, the first continuous mill was in production. The rolling of wide strips several hundred feet in length was the achievement of H M Naugle and A J Townsend working at the Columbia Steel Co. in Butler, Pennsylvania, USA.

Electrolux absorption refrigeration system
Carl Munters and Baltzar von Platen, undergraduates at the Royal

Institute of Technology in Stockholm, Sweden. The patent was purchased by Dr Alex Wenner Gren of AB Electrolux, Sweden.

Film (3-dimensional motion pictures)
Perfect Pictures, US. Viewers were given spectacles with one red and one green lens to watch 'The Power of Love'.

Flettner rotorship
Anton Flettner of the University of Göttingen, Germany. It derived its power from the lift and drag forces developed in large, rotating cylinders mounted above the hull. In 1925 a Flettner rotorship, 'Buckau', 680 tons, crossed the Atlantic Ocean.

Football pools
John Jervis Barnard, ex-Coldstream Guards officer, in Birmingham, UK. Called the 'Pari-Mutual Pools'.

High-frequency induction furnace (steel)
Dr Edwin Fitch Northrup, of both Princetown University and the Ajax Electrothermic Corp., Trenton, New Jersey, USA.

Infra-red grill (sandwich toaster)
UK.

Motorised perambulator
Dunkleys of Birmingham, UK.

Patterned concrete building blocks
Frank Lloyd Wright, US architect.

Phase shift distortion (radio waves)
Dr Hoyt Taylor and Leo C Young, US Navy scientists.

Radio advert
WEAF Station. 10 minutes were sold to a furniture promoter of Jackson Heights, New Jersey, USA.

Radio-broadcasting station (UK)
a Marconi team at their laboratory in Writtle, near Chelmsford, with weekly half-hour broadcasts.

Self-winding wrist-watch
John Harwood, UK watchmaker and repairer. Unable to interest Swiss manufacturers, Harwood and Cutts formed a company which employed Swiss firms to make their watch. Between 1928 and 1931, 30,000 Harwood self-winders were made before the business was forced into liquidation.

Sky-writing (smoke-trails)
Major J C Savage of Hendon, London, using SE5 aeroplanes aerobatically to write messages in the sky.

Tank (revolving turret)
Vickers Medium Mark I; UK.

Theory of polymers
Hermann Staudinger, German chemist at the Federal Institute of Technology, Zurich.

Tractor (diesel-engined)
Benz of Stuttgart, Germany. A two-cylinder, three-wheel machine was soon followed by a four-wheel version.

Tri-Ergon sound film
Joseph Engl, Joseph Massolle and Hans Vogt. The first film, based on a Hans Andersen tale, was made by UFA, Germany's leading film company.

Tubular immersion heater unit (kettle)
Arthur Leslie Large; manufactured by Bulpitt & Sons Ltd of Birmingham, UK.

Vitamin E
Herbert McLean Evans and K S Bishop, physiologists, US.

Water-skiing (powered)
Ralph W Samuelson at Lake City, Minnesota, USA. By 1925 he was able to go over a ski ramp lubricated with cooking lard, and also took tows from Curtiss flying-boats.

1923
Autogyro
Juan de la Cierva. Following early and largely unsuccessful experiments in his native Spain, from 1926 the Cierva Autogyro Co., privately financed by Lord Weir, was based in the UK. Cierva was killed in a commercial air-crash in 1936.

Coding machine (production model)
German Cipher Machines Co. *It was called the 'Enigma', as developed by* Dr Arthur Scherbius, *engineer, Berlin.*

Compton effect
Arthur Holly Compton, *Professor of Physics, Chicago University,* USA.

Frozen food
Clarence Birdseye, *former field naturalist and fur trader,* USA, *who established a company to industrialise his process in 1924 at Gloucester, Massachusetts. The first retail sales of packaged frozen peas, etc. were at Springfield in 1930. Pre-cooked frozen foods were introduced nine years later.*

Hafnium
Georg von Hevesy, *Hungarian-Swedish chemist at the Institute of Physics, Copenhagen, with* Dirk Coster, *Dutch physicist.*

Hearing aid (valve system)
Marconi Ltd, *London. Called the 'Otophone', it weighed 16 lb.*

Hectographic (or spirit) duplicating
Ormig Gesellschaft, *Germany.*

Iconoscope (electronic camera tube)
Vladimir Kosma Zworykin, *Russian émigré to the* USA. *He also developed his picture tube display, the kinescope, during the same year.*

In-flight refuelling (aircraft-to-aircraft)
US Army Air Service *with de Havilland four-day bombers.*

Lorry (diesel-engined)
Benz *of Stuttgart, Germany.*

Microphone (ribbon-type)
W H Schottky *and* Erwin Gerlach, *Germany.*

Low-pressure automobile tyre
Simultaneously by the Firestone Tire & Rubber Co., Akron, Ohio, USA, *as the 'Balloon'; and by Michelin et Cie, Clermont-Ferrand, France, as the 'Confort'.*

Phonofilm
Professor Lee de Forest, US, *who showed a number of singing and musical shorts using his process, at the Rialto Theater, New York.*

Planetarium (geocentric type)
Dr Walther Bauersfield *of the Carl Zeiss Optical Company, Jena, Germany, for the German Museum in Munich.*

Sound film on motion pictures (experimental):
Joseph Tykocinsky Tycocyner, *University of Illinois,* USA.

Short-wave radio transmitter
Dr Frank Conrad *of Westinghouse Electric Corp. from the* KDKA *radio station in Pittsburgh, Pennsylvania to* KDPM *radio station, 100 miles away in Cleveland, Ohio.*

Window polish
Reckitt & Sons Ltd, *Hull,* UK – *'Windolene'.*

Whooping cough vaccine
Thorvald Madsen, *Danish bacteriologist.*

1924

Aerial archaeology
O G S Crawford, *former* RFC/RAF *observer, was sponsored by A Keiller to fly over Hampshire, Wiltshire and Dorset,* UK. *The results were published in a book, 'Wessex from the Air'.*

Aetherophon
Professor Thérémin *of Leningrad,* USSR: *an electric musical instrument where the player regulated the pitch simply by movements of the hand towards and away from an upright rod connected with a valve.*

Aga kitchen range
Nils Gustav Dalén, *blind Swedish physicist.*

Automatic post office
installed in Bath, UK. *The kiosk combined posting box, stamp-vending machine and pay phone.*

'Beetle' (thermosetting plastic)
Fritz Pollak, *Vienna, Austria. Instead of heating phenol-formaldehyde resin*

under pressure, he heated urea. Unlike Bakelite, Beetle was odourless and tasteless, and much brightly coloured tableware came to be made in this material.

Cellulose paint finish
Oakland Cars, US.

Chanel No. 5 (perfume)
Mademoiselle 'Coco' Chanel, *Paris.*

Disposable paper handkerchiefs
Kimberley-Clark Co. *of Nennah, Wisconsin,* USA: *'Kleenex'.*

Fast rotary press (punched cards)
International Business Machines, US.

Gas chamber (execution)
Major D A Turner, US *Army Medical Corps. It was used at Nevada State Prison, Carson City, for the execution of a convicted murderer, Gee Jon. He took 6 minutes to die.*

'Harpic' (lavatory cleaning agent)
Harry Pickup, *Harpic Manufacturing Co., London.*

International chess match by radio
Between Haverford College, Philadelphia, USA, *and Oxford University,* UK. *Facilitated by five and a half hours' clear transmission.*

Language laboratory (gramophone-based)
Ralph Waltz *at the University of Utah,* USA, *was the first to use this phrase.*

Motorway
Italian Autostrada between Milan and Varese.

Pauli exclusion principle
Wolfgang Ernst Pauli, *Austrian-Swiss theoretical physicist, Federal Institute of Technology, Zurich.*

Pilotless aircraft flight
N-9 aircraft, as equipped by the US *Naval Research Laboratory, College of Ordnance and Naval Proving Grounds.*

'Presdwood' (prototype hardboard)
William Horatio Mason *in a sawmill*

shed in Laurel, Mississippi, USA, *while attempting to make a by-product out of sawmill waste, using a self-designed steam press. He began a factory in 1926 and trade-named his board 'Masonite'. Through one of his associates, Arne Asplund, the technology spread to Sweden in 1929.*

Undulatory theory of matter
Prince Louis Victor Pierre Raymond de Broglie, *French physicist, whilst investigating electron beams at La Sorbonne.*

Variable-pitch airscrew
Dr H S Hele-Shaw *and* T E Beacham, UK. *After convincing tests, the British Air Ministry ordered a dozen of these airscrews, but there was a general lack of interest until 1937 when 'Rotol' Airscrews Ltd began serious production.*

Variable-propellor water turbine
Viktor Kaplan, *Austrian engineer at Brunn, Czechoslovakia.*

Wash/spin-dry machine (electric-powered)
Savage Arms Corp., *New York,* USA.

1925

Aerial crop-dusting service (commercial)
C E Woolman *of Huff-Daland Dusters Inc., flying a Petrel aircraft to discharge calcium arsenate over a Georgia cotton plantation.*

Bélinographe
Édouard Belin, *French researcher.*

Cinemascope
Dr Henri Chrétien, *French physicist: wide-screen cinema.*

Continuously variable friction gear
Frank Hayes *of New Jersey,* USA.

Electrodynamic loudspeaker (high-fidelity)
C. W. Rice *and* E. W. Kellogg, *General Electric Co., Schenectady, New York,* USA.

Gramophone (all-electric)
Brunswick Co. *of Dubuque, Iowa,* USA

– *called the 'Brunswick Panatrope'.*

Masurium and rhenium
Ida Eva Noddack *and her husband,*
Walter Karl Friedrich Noddack,
*chemists, National Physical
Laboratory, Germany.*

Motor car chassis (all-steel)
André Citroën, *France: the Citroën
B10 and B12.*

Petrol pump clock dial
S F Bowser and Co., *Fort Wayne,
Indiana,* USA.

**'Pulse ranging' (ionospheric
measurement)**
Dr Gregory Breit *and* Dr Merle A
Tuve, *Carnegie Institution of
Washington* DC, USA.

Quantum mechanics
Max Born, *Professor of Physics,
Göttingen, Germany, and* Werner
Karl Heisenberg, *German theoretical
physicist in Copenhagen, Denmark.*

Rigid airship (helium-filled)
US Navy, *Lakehurst, New Jersey,* USA.
It was designated the Z-R1.

Teaching machine
Sidney L Pressey *of Ohio State
University,* USA.

1926
Aerosol can
Erik Rotheim, *Norway.*
Automatic traffic lights
London.
Bulldozer
Robert Gilmour Le Tourneau,
*muck-shifting contractor, while
building a highway between Oakland
and Stockton, California,* USA. *He
fitted a steel scraper blade in front of a
Best tractor and equipped it with an
electric cable.*
'Buna S' (synthetic rubber)
I G Farben, *Germany.*
Cellophane (moisture-proof)
William Hale Charch *and* Karl
Edwin Prindle *of the Du Pont
Cellophane Co.,* USA.
**Crossbar (electro-mechanical
telephone exchange)**
Sweden.

**Electro-mechanical television
systems**
John Logie Baird, *Scottish inventor
in London and* Charles Francis
Jenkins *in Washington and* D Mihaly,
*Hungarian inventor in Munich. They
used neon lamps, Nipkow discs and
photoelectric cells.*
**Lawn-mower (electric-mains-
powered).**
Liquid-fuelled rocket
Robert H Goddard, *Professor of
Physics at Clark University,
Massachusetts,* USA. *It was fired to 41
ft altitude from nearby Auburn.*

Goddard beside his liquid-fuelled rocket

Macromolecular chemistry
Hermann Staudinger, *Professor of
Organic Chemistry at Freiburg,
Germany.*
Power steering (motor vehicles)
Francis W Davis, *former chief
engineer of the Pierce-Arrow Motor
Car Co., at a small engineering shop in
Waltham, Massachusetts, with the
assistance of* G W Jessup, *tool-maker*

and engineer.
Propeller-rudder (marine)
E Schneider *of Vienna. It was first constructed by Messrs J M Voith of Heidenheim for an experimental launch, 'Torqueo'.*
Technicolor (cinematography)
Herbert Thomas Kalmus, D F Comstock *and* W B Wescott *at the Massachusetts Institute of Technology, Cambridge,* USA.
Tungsten carbide (industrial quantities)
Krupp Gesellschaft, *Germany, as* 'Widia', *and* General Motors, US, *as* 'Carboloy'. *It was originally discovered twenty years before by* Henri Moissan, *French chemist.*
Urease (crystalline enzyme)
James Batcheller Sumner, *biochemist, Cornell University Medical School,* USA.
Waterproof watch
Rolex, *Switzerland; called the* 'Oyster'.

1927
Amyl cinnamic aldehyde (synthetic perfume)
having the scent of jasmine.
Continuous casting of metals
Dr Siegfried Junghans, *at the brassworks he ran for his family firm of clock and watchmakers in Germany. He later progressed to other metals, including steel, and his process was commercially exploited by* Irving Rossi *of New York.*
Electron diffraction by crystals
simultaneously by Clinton Joseph Davisson *with* L H Germer *at Bell Telephone Laboratories, and* George Paget Thomson, *Professor of Physics at Aberdeen,* UK.
Ethylene glycol
'anti-freeze' liquid for motor car radiators.
Feature-length sound film
Warner Brothers *of Hollywood, California,* USA. *Called 'The Jazz Singer', it starred Al Jolson.*

Frogman flippers
Louis de Corlieu, *France.*
Genoa jib (sail)
Sven Salen, *Swedish helmsman, whose boat 'Lilian' had a 'balloon staysail' at the International 6-Metre Class races of the Genoa Regatta, Italy.*
Iron lung (medical)
Professor Philip Drinker *of Harvard University. It was manufactured by Warren E Collins Inc. of Boston, Massachusetts as the 'Drinker Respirator'.*
Jukebox (all-electric)
Automatic Musical Instrument Co. *of Grand Rapids, Michigan, and* Seeburg Co. *of Chicago, Illinois,* USA.
MNS blood groups
Karl Landsteiner *and* Philip Levine *at the Rockefeller Institute of Medical Research, New York,* USA. *They also discovered the 'P' system the same year.*
Motorboat (rocket-powered)
Steel hull by Mullins Manufacturing Corp., *Salem, Ohio,* USA. *Modified in Florida by Dick Pope and test-piloted by his brother Malcolm, 'Pirate Kid' was powered by 32 rockets fired from its iron-plated hull, and went out of control at 60 mph when its rudder burnt through.*
Motor car bodywork cleaner
Reckitt & Sons Ltd, *Hull,* UK; *called* 'Karpol'.
Pentode (five-electrode radio valve)
Dr Gilles Holst *and* Bernard Tellegen, *Philips Research Laboratories, Eindhoven, Netherlands – called the 'B443'.*
Pop-up toaster
Charles Strite, *mechanic from Stillwater, Minnesota,* USA – *called the* 'Toastmaster'.
Rocket-propelled car
Fritz von Opel, *Germany.* RAK *1 was powered by two solid-fuel rockets and fitted with aerofoils.* RAK *2 achieved 195 mph at the Avus Circuit, Berlin.*

Salad cream
H J Heinz Co. Ltd, *Harlesden,
London,* UK.
Sex hormone
Bernhard Zondek *and* Selmar
Ascheim.

1928

Catalytic cracking of petroleum
Eugene J Houdry, *French engineer,
whose search for a catalyst that could
be regenerated led him to examine over
1000 substances before he discovered
activated clay. His process was
perfected and made practical first by
Vacuum Oil Co., then Sun Oil Co.*
Cloverleaf intersection
*State Highway Routes 4 and 25,
Woodbridge, New Jersey,* USA.
**Colour television transmission
(electro-mechanical)**
John Logie Baird *at the Baird
Studios, Long Acre, London.*
The Diels-Alder reaction
Otto Paul Hermann Diels, *Professor
of Organic Chemistry, Kiel, Germany,
with his assistant* Kurt Alder.
'Elastoplast' (adhesive bandage)
T J Smith & Nephew Ltd, *Hull,* UK.
**Four-cylinder, two-stroke
outboard engine**
Ole Evinrude, US – *called the 'Elto
Quad'.*
Geiger counter
Hans Geiger, *Professor of Physics at
Tübingen University, and his assistant*
E W Müller, *for the purposes of
investigating Beta-ray radioactivity
in atoms. The Geiger-Müller tube is
named after them.*
Hot-air hand-drying apparatus
Elektra Händetrockner GmbH,
Germany.
'I Speak Your Weight' machine
Joseph Tripodi. *It was first installed
in Genoa, Italy.*
Image dissector tube (television)
Philo Taylor Farnsworth, *engineer,
Utah,* USA.
Ondes musicales
Maurice Martenot, *French composer
in Paris. The electric musical
instrument was played using a ring on
the finger fastened to a cord, producing
a wailing glissando.*
Pedal wireless
Alfred Traeger, *engineer, to assist*
John Flynn *in his Flying Doctor
Service in Australia.*
Penicillin
Alexander Fleming, *Scottish
bacteriologist, St Mary's Hospital,
Paddington, London.*
PVC (poly-vinyl-chloride)
simultaneously by Carbide and
Carbon Chemical Corp. *and* Du
Pont, *both* US, *and* I G Farben,
Germany.
The Raman effect
Chandrasekhara Venkata Raman,
physicist, Calcutta University, India.
Robot
Captain Rickards *and* A H Reffell *at
Gomshall, Surrey,* UK. *It opened the
Model Engineering Exhibition,
London.*
Synthetic rubber (commercial)
J C Patrick. *Marketed as 'Thiokol A',
it had remarkable oil-resistance, but its
very unpleasant smell hampered its
success.*
'Talking beacon' (lighthouse aid)
Charles Alexander Stevenson.
Teletype
Edward Ernst Kleinschmidt *of
Morkrum-Kleinschmidt Co., Chicago,*
USA.
Television recording system
John Logie Baird, *Long Acre,
London. His 'Panovision' used
aluminium gramophone records.*
Tomography
André Bocage, *France.*
Tubular-nibbed fountain pen
Rotring ('red ring'), *Hamburg,
Germany. It was called the
'Tintenkuli'.*
**Vitamin C (isolated from citrus
fruits)**
Albert von Nagyrapolt Szent-
Györgi, *Hungarian biochemist at
Cambridge University,* UK.

1929

Airship (diesel-engined)
Lieutenant-Colonel Victor
Richmond, *leading a design team at*
Cardington, Bedford, UK. *The five-*
Beardmore-engined R101 *airship*
crashed at Beauvais, France, in 1930.

Car radio (commercial
manufacture)
Galvin Manufacturing Corp.,
Chicago, USA. *It was called the*
'Motorola'.

Coaxial cable
Bell Telephone Laboratories, US.

Electrical-mechanical flight
simulator
Edwin Albert Link, US. *It was*
adopted by the US *Navy in* 1931.

Electro-encephalogram (EEG)
Hans Berger, *Professor of Neuro-*
psychiatry at Jena. It was popularised
five years later by Nobel prize-winner
Lord Edgar Douglas Adrian, UK.

Electron microscope
Max Knoll *and* Ernst Ruska *in*
Berlin. Three years earlier they had
set out to investigate the discovery
made by Dr Hans Busch *of Jena*
University that when a beam of
electrons passes through a wire-coil,
which acts as a magnet, the beam can
be focused by a lens. Knoll and Ruska
used a coil to form an image of a small
aperture at a magnification of one;
they then added a second stage which
stepped up the magnification to
seventeen. By 1933 *Ruska's*
'Supermicroscope', in achieving
magnifications of up to 12,000, *had*
gone well beyond the resolution of the
optical microscope.

Esterone
Adolf Friedrich Johann Butenandt,
German biochemist at Göttingen,
simultaneously with and independently
of Edward Albert Doisy, *Director of*
the Department of Biochemistry at St
Mary's Hospital, St Louis, Missouri,
USA.

Foam rubber
Dunlop Latex Development
Laboratories, *Fort Dunlop,*
Birmingham, UK. *A Dunlop scientist,*
E A Murphy, *had the idea of beating*
latex into a foam with an egg-whisk.
His colleague W H Chapman *used a*
gelling process to produce a rubber
foam capable of setting in moulds.
'Dunlopillo' was soon being used for all
forms of upholstery.

Giant flying-boat (twelve
engines)
Dornier *of Germany. The* 150-
passenger DoX just managed a
transatlantic crossing, but otherwise
proved totally unmanageable.

Hydroponics (growing plants in
water)
William Frederick Gericke,
University of California, USA.

Magnetic sound-recording tape
(plastic)
Dr Fritz Pfleumer, *Austrian research*
worker with I G Farben, *Allgemeine*
Electrizitäts Gesellschaft. Cellulose
acetate tape was later replaced by
*'Luvitherm' (*PVC*) tape, running at* 30
inches per second.

Oxygen-lance steel-making
process
Dr Robert Durrer *and his assistant*
Dr Heinrich Hellbrügge *at the Berlin*
Institute of Technology. The first
experiments were at von Roll AG,
Switzerland in 1948. *A full-scale*
operation was perfected by Dr
Theodro Eduard Suess *at the*
Austrian steelworks, VOEST.

Quartz crystal clock
Warren Alvin Marrison, *clockmaker*
of Orange, New Jersey, USA.

Radiotelephone service
set up on the steamship 'Leviathan'.

Retractable undercarriage
(hydraulically operated)
George Messier, *France.*

'Scotch' Tape
Richard Drew, *research technician for*
the Minnesota Mining and
Manufacturing Corp. (3M*), at St*
Paul. It was introduced to Britain
from France as 'Sellotape'.

Submersible decompression chamber
Sir Robert H Davis *of Siebe-Gorman Ltd. It was successfully tested in Loch Long, Scotland.*

Synchromesh gearbox
Cadillac *and* La Salle, USA.

Synthesis of haemin
Hans Fischer, *organic chemist, Munich Technical High School, Germany.*

Television transmitting studio
John Logie Baird *at Long Acre, London. Sound and vision were transmitted in 1930 to the owners of his 'Televisor' receiving apparatus.*

Toughened plate-glass windscreen
St Gobain, *French glassmakers, based on discoveries by the* US *Corning Glass Company.*

Tune-playing motor horn
Sparks & Witherington Co.

1930
Bathysphere
Charles William Beebe *and* Otis Barton, US. *In 1934 Beebe recorded a depth of 3000 ft in their hollow steel globe off Bermuda.*

Cosmic ray magnetic cloud chamber
Carl David Anderson *and* Robert Andrews Millikan, *California Institute of Technology*, USA.

Cyclotron
Professor Ernest O Lawrence *and* N E Edlefsen *at the University of California, as a sophisticated particle accelerator to study the internal structure of disintegrated atoms. Lawrence had been inspired by a paper written by a German physicist called Wideroe.*

Differential analyser
Dr Vannevar Bush *and a team at the Massachusetts Institute of Technology, Cambridge*, US. *It was mechanically operated. Mark II was produced in 1942.*

Francium
Fred Allison, *physicist, Alabama Polytechnic Institute*, USA, *using his magneto-optic method.*

Hygienic tampon
Earl Hass, US. *His 'Tampax' Co. was formed in 1937.*

Ion-radiated food preservation
Wust, *France.*

Pepsin (crystalline enzyme)
John Howard Northrop, *chemist, Rockefeller Institute of Medical Research, New York*, USA.

Photo-flash bulbs (for photography)
General Electric Co., Schenectady, New York, USA.

Pluto (planet)
Clyde Tombaugh, *astronomer, at the Lowell Observatory, Flagstaff, Arizona*, USA, *using photographic methods.*

Polystyrene
I G Farben, *Germany, from benzine or crude oil. One of the first commercial products was 'Styron', marketed in 1937.*

Railway carriage air-conditioning
B & O Railway Co., USA. *For its dining-car 'Martha Washington', as tested behind 'The Columbian' between Baltimore and Cumberland*, USA.

Radiographic enlarger
Irving Langmuir, *physical chemist at General Electric's research laboratories, Schenectady*, USA.

Radio-pulse detection of ships (bounce-back system)
W A S Butement *and* P E Pollard, *Signals Experimental Establishment, Woolwich, London.*

Self-regulating hydraulic shock absorber
F G G Armstrong.

Supermarket
Michael Cullen *of Long Island, New York. It was called the 'King Cullen Food Stores'.*

Telescopic umbrella
Hans Haupt, *Berlin, Germany.*

Televisual telephone
'Ikonophone' – American Telephone
& Telegraph Co. USA. A two-way
telephone with two-way television
transmission.

1931
'Alka-Seltzer'
Miles of Elkart, Indiana, USA.
Amphibious bicycle (inflatable floats)
Alfred Kreutz, forester from Sybba,
East Prussia.
Amphibious tank
Vickers-Armstrong, UK. The
Vickers-Carden Lloyd A4 was
powered by a 50 hp Meadows engine.
Androsterone
Adolf Friedrich Johann Butenandt,
privatdozent chemist, Göttingen,
Germany.
Astatine
Fred Allison, Professor of Physics,
Alabama Polytechnic Institute, USA.
'Blood bank'
Professor Sergei Sergeivitch Yukin
at the Sklifosovsky Institute,
Moscow's central emergency service
hospital. The term 'blood bank' was
coined six years later by Bernard
Fantus at Cook County Hospital,
Chicago, USA.
Caterpillar tractor (diesel-engined)
Caterpillar Tractor Co., Peoria,
Illinois, USA.
Commercial electric guitar
Adolph Rickenbacker, Los Angeles-
based machine and engineering
contractor, with Messrs Beauchamp
and Barth. Their Electro String
Instrument Corp. produced an
aluminium and steel banjo nicknamed
the 'Frying Pan'.
Electric-generator windmill
G Darrieus, France. Three sails of
aerofoil section like the blades of an
aeroplane propeller were mounted on a
pylon.
Electric kettle (resettable, self-ejecting safety plug)

Walter Henry Bulpitt of
Birmingham, UK.
Empire State Building
New York City, USA. Messrs Shreve,
Lamb and Harman, architects with
N. G. Balcom Associates, engineers. It
had 102 storeys and was 1250 ft high.
Exposure meter (photoelectric)
J Thomas Rhamstine of Detroit,
Michigan. It was called the
'Electrophot'.
Glass-fibre
Owens Illinois Glass Company, US
bottlemakers. Glass-fibre cloth was
manufactured for heat and cold
insulation. The Owens-Corning
Fiberglass Corp. was formed in 1938.
Guitar pick-up
Horace Rowe, engineer, and De
Armond, musician, US.
Iconoscope electronic TV camera
Vladimir Zworykin at the
Westinghouse laboratories, Pittsburgh,
Pennsylvania, USA.
Incompleteness Theorem
Kurt Gödel, German logician.
Microphone (crystal type)
C B Sawyer of Cleveland Heights,
Ohio, USA.
Neutrino
Wolfgang Pauli, Austrian Professor of
Physics, Zurich, Switzerland. Not
until 1956 was a neutrino beam
detected by F Reines and C L Cowan,
Los Alamos, New Mexico, USA.
Radio astronomy
Karl Guthe Jansky, radio engineer at
Bell Telephone Laboratories,
Holmdel, New Jersey, USA. Using an
improvised aerial, Jansky picked up
radio emissions – as opposed to
traditional light rays – from the Milky
Way.
Railcar (pneumatic-tyred)
Edouard Michelin, tyre
manufacturer, with Jacques Hauvette
and Pierre-Marcel Bourdon, at the
Michelin factory, Clermont-Ferrand,
France. Called the 'Micheline'; by
1939, 140 such railcars were in use

throughout Europe.
Razor (electric)
Colonel Jacob Schick, US. *It was
manufactured at Stamford,
Connecticut*, USA.
Tailless aircraft
Dr Alexander Martin Lippisch,
aerodynamicist, Germany.
**TWX (teletypewriter exchange
service)**
Bell Telephone & Telegraphy
Company, US. *The British Post
Office's Telex system was operative by
1932.*
**Tractor (rubber-tyred
production model)**
B F Goodrich, *Akron, Ohio*, USA.

1932
Car radio
Blaupunkt *five-valve AS5. It was
first fitted in a* US *Studebaker
automobile.*
Defibrillator
Dr William Bennett Kouwenhoven,
*biomedical engineer, Dean of the
Engineering school, Johns Hopkins
University, Maryland*, USA.
**Deuterium (heavy hydrogen
isotope)**
Harold Clayton Urey, *chemist,
Columbia University*, USA.
Distant-reading compass
L C Bygrave, *Royal Aeronautical
Establishment, Farnborough*, UK.
Gallup Poll
Dr George Horace Gallup, *Professor
of Journalism at Northwestern
University, Iowa*, USA.
**Infra-red medical photography
(subcutaneous analysis)**
Professor Haxthausen *of
Copenhagen, Denmark.*
Kort nozzle rudder (marine)
Ludwig Kort, *German
aerodynamicist.*
**Lethane (synthetic organic
insecticide).**
**Lithium disintegration (proton
bombardment)**
John Douglas Cockcroft *and* Ernest

Thomas Sinton Walton, *nuclear
physicists, Cavendish Laboratory,
Cambridge*, UK.
**'Mars Bar' (nougat and
caramel-filled chocolate bar)**
Forrest Mars, *Mars Confection Ltd,
Slough*, UK.
**Motion pictures (stereophonic
sound)**
Abel Gance *and* André Debrie, *film-
makers, Paris. They made a re-edited
version of Gance's 1927 eight-hour
silent epic, 'Napoléon Bonaparte'.*
Neoprene
Arnold M Collins *and* Dr Wallace H
Carothers *of Du Pont*, USA, *based on
the twenty-year pioneer research of a
religious Belgian-born pure scientist,
Father Julius A Nieuwland and his
'divin-acetylene'.*
Neutron
James Chadwick, *physicist,
Cavendish Laboratory, Cambridge*,
UK.

*Dr Wallace Carothers, inventor of Du
Pont's nylon and neoprene*

Parking meter
Carlton C Magee, *editor of 'Oklahoma City Newspaper' and chairman of the Businessmen's Traffic Committee in that city. He further refined his coin-operated machine three years later, as manufactured by his Dual Parking Meter Co.*

Positron
simultaneously by Carl David Anderson, *California Institute of Technology,* USA, *and* Patrick Maynard Stuart Blackett, *Cavendish Laboratory, Cambridge,* UK.

Synthesised Vitamin C
Walter Norman Haworth, *Professor of Organic Chemistry at Birmingham University,* UK.

Thiram (dithiocarbonate fungicide)
Du Pont de Nemours, USA.

VHF radio-telephony
Marquese Guglielmo Marconi, G A Mathieu *and* G A Isted, UK.

Tractor tyre (pneumatic, low-pressure)
Firestone Tire & Rubber Co., Akron, Ohio, USA. *First tested by Harvey S. Firestone on his farm in Columbiana.*

Wind tunnel (for mass-production automobile design)
Ford Motor Co., *Detroit, Michigan,* USA *– for their Lincoln car.*

Zoom lens (cinematography)
Taylor Taylor & Hobson *of Leicester,* UK, *with their 'Cooke' lens; then* Bell & Howell, *Chicago,* USA. *The first practical zoom lens was produced 21 years later by* Angenieux, *France.*

1933

Anchor (CQR)
Geoffrey I Taylor, *Cambridge,* UK. *To enable flying boats/seaplanes to moor on the water.*

Bicycle (supine-recumbent)
Charles Mochet, *France, ridden by François Faure and called the 'Vélocar'.*

'Dayglo' fluorescent pigments
Joe *and* Bob Switzer *at their father's drug store in Los Angeles, California,* USA. *Together with chemist* Dick Ward, *the Switzers began a lithography company in Cleveland, Ohio to produce fluorescent inks such as 'Dayglo Orange'.*

'Dettol' (antiseptic liquid)
Reckitt & Sons Ltd, *Dansom Lane, Hull,* UK.

Flue-gas desulphurisation process
London Power Company's research chemists, *to eliminate chimney gases from the newly built Battersea 'A' power station. It was known as 'lime-scrubbing'.*

Frequency Modulation (FM)
Edwin Howard Armstrong, *Professor of Electrical Engineering, Columbia University,* USA.

Microscope (miniature, hand-held and versatile)
John McArthur, *medical student and Leica camera enthusiast, University College Hospital, London. Although a dozen instruments were manufactured in 1936, it was not until 1959 that Cooke, Troughton & Simms Ltd put it into series production.*

Million-volt electrostatic particle accelerator
Merle A Tuve, *at the Department of Terrestrial Magnetism of the Carnegie Institution of Washington,* USA. *It was based on the work of* Robert van der Graaff, *former Rhodes scholar from Alabama,* USA.

'Monopoly' (board game)
Charles Darrow, US.

Multiple-flash sports photography
Harold E Edgerton *and* Kenneth J Germeshausen, *Massachusetts Institute of Technology, Cambridge,* USA.

Non-magnetic engine
Vickers-Petter Ltd, UK, *for 'Research', a motorboat to explore the earth's magnetic field. The Atomic 160 hp engine was mainly of bronze alloy.*

Pantothenic acid
Roger J Williams, *biochemist*, US, *et al.*

Polythene (polyethylene)
Dr Reginald Gibson *and* E W Fawcett *at* ICI*'s Alkali Division research laboratories, Winnington, Cheshire,* UK. *The first ton of polythene was complete in December 1938, and a factory went into production less than a year later.*

Sodium vapour lamp
General Electric Co., Schenectady, New York, USA.

Stereophonic disc recording
Alan Dower Blumlein, UK *scientist working for Electric and Musical Industries (later* EMI*), London.*

Technicolour cartoon film
Walter Elias Disney, *producer*, US, *whose 'Flowers and Trees' achieved a tremendous success and was soon followed by 'Three Little Pigs'.*

Telephone speaking clock
M Esclangon, *Director of the Paris Observatory, for the French Ministry of Posts and Telecommunications.*

Tractor (three-point hydraulic plough hitching system)
Harry George Ferguson, *engineer, Northern Ireland.*

Vitamin B2 (riboflavin)
Richard Kuhn, *Austrian chemist*, Albert von Nagyrapolt Szent-Györgyi, *Hungarian biochemist, and* Julius Wagner von Jauregg, *Austrian neurologist. Synthesised 2 years later.*

1934
Automatic injection moulding press
for the various acetate and formaldehyde moulding powders recently developed.

Catseye
Percy Shaw, *road-repairer, Halifax, Yorkshire,* UK – *a reflecting road-stud to enable motorists to see in dark or foggy conditions. Shaw was particularly inspired by the glinting eyes of a cat sitting on a fence.*

Cerenkov Effect
Pavel Alekseevich Cerenkov, *physicist, Lebedev Physical Institute, Moscow. It was further developed during the next three years by* Ilja Michajlovic Frank *and* Igor Jevgenevic Tamm, *also of the Lebedev Institute,* USSR.

Colt chimney cowl
Herr Kuckuck, *to prevent downdraught and smoky fires.*

Diesel-electric railway traction
Charles Franklin Kettering, *General Motors Corporation, Detroit,* USA, *with* George Codrington (*ex-Winton Gas Engine Co.) and* Harold L Hamilton (*ex-Electromotive Company). The first* US *diesel-powered main line locomotive, 'General Motors 103', made its first 83,000-mile test run in 1939.*

Drive-in cinema
Richard Hollingshead, *on a 10-acre site at* Camden, *New Jersey,* USA.

Electromagnetic detection of aircraft (experimental)
Pierre David *at Le Bourget Airport, France.*

Launderette
J F Cantrell, *Fort Worth, Texas,* USA. *It was called the 'Washeteria'.*

Oxidation enzymes
Axel Hugo Theodor Theorell, *crippled Swedish biochemist, Kaiser Wilhelm Institute, Berlin.*

'Perspex' (methyl methacrylate polymer)
Dr Rowland Hill *and* researchers at Imperial Chemical Industries, UK. *Similar researches by* Dr Otto Rohm *and his colleague* Hass *of Darmstadt, Germany, produced 'Plexiglas' and by* Du Pont de Nemours, USA, *'Pontalite' then 'Lucite'.*

Progesterone (hormone)
Adolf Friedrich Johann Butenandt, *Professor of Organic Chemistry, Danzig Institute of Technology, Germany.*

Radio beam navigation aid
Dr Hans Plendl, *high-frequency radio*

expert from Rechlin, Germany. Called *'Knickebein' (bent leg)*, it was first used by the Luftwaffe in 1940.
Radio-synthetic organ
Abbé Pujet *in the Church of Notre Dame de Liban of Paris. The harmonics of the pipes were picked up by microphones, mixed synthetically into new tone colours and finally sounded over loudspeakers.*
Streamlined production automobile
Chrysler 'Air-Flow' Coupé, USA, *and the 'Traction Avant' Citroën, France.*
Synthesised prontosil dye
Dr Gerhard Domagk, *pharmacologist employed by I G Farben, Germany.* Daniel Bovet, *Swiss-born scientist, identified prontosil dye's active compound as sulphanilamide. Sulphonamides became the latest generation of chemotherapeutic drugs.*
Tannoy portable loudhailer
Guy R Fountain *of the Tulsmere Manufacturing Co.*, USA. *The name was based on the words 'tantalum alloy'.*
Tone-wheel electric organ
Laurens Hammond *and* John Hanert, *engineers of Chicago, Illinois,* USA, *using the synchronous motor Hammond had developed for a non-ticking clock. The public début of the 'Hammond Model A' came in 1939 at the Industrial Arts Exposition in New York's Radio City* RCA *Building.*
Vitamin K
Carl Peter Henrik Dam, *Professor of Biochemistry at Copenhagen University, Denmark, simultaneously with* Edward Adelbert Doisy, *Director of the Department of Biochemistry at St Mary's Hospital, St Louis, Missouri,* USA.

1935
Aero-advertising
Richard Ormonde Shuttleworth, *wealthy young aviator of Old Warden, Bedfordshire,* UK. *He experimented with fitting a series of neon letters not* exceeding 4 ft to the underwings of his de Havilland 84 Dragon biplane. BEER IS BEST *or* OXO *would appear in the night sky. Although a company was formed, the project failed when aircraft were banned from flying below a certain level.*
Aircraft (geodesic construction)
Barnes Neville Wallis, *designer and engineer, Vickers Aviation Ltd, Weybridge, Surrey,* UK – *the Vickers G4/31 monoplane.*
Colour film (monopak-still)
Leo Godowsky Jr *and* Leopold Mannes, *two music students and self-taught chemists in New York,* USA. *Following years of initial experiments in Mannes' kitchen, the two Leos were later employed by Dr C E K Mees of Eastman-Kodak and 'Kodachrome' was the commercial result. The following year the German 'Agfacolor' was marketed, also as a monopak-process.*
Dellinger Effect (sunspots on radio communication)
John Howard Dellinger, *physicist, Chief of the Radio Section,* US *National Bureau of Standards.*
Desk calculator (tabulator with independent cycles)
Compagnie des Machines Bull, *Paris, France.*
Electric typewriter (mass-produced)
International Business Machines, USA.
Fireboat (steel-hulled and diesel-engined)
J Samuel White Ltd, *Cowes, Isle of Wight,* UK. *The 78 ft 'Massey Shaw' for work on the River Thames, London.*
Fluorescent lighting (hot-cathode)
General Electric Co. *at Nela Park, Ohio,* USA.
Hearing aid (electronic)
Edwin A Steven's *'Amplivox'.*
Ink cartridge (pen)
M Perraud, *Director of Jiff-*

Waterman, France.
Mesotron
Hideki Yukawa, *physicist, Osaka University, Japan. Two years later* S Neddermeyer, J Street, E Stevenson *and* C Anderson *verified Yukawa's postulation about what later became known as the meson.*
Microphone (non-directional)
Bell Telephone Laboratories, Holmdel, New Jersey, USA. *Called the 'Western Electric Eight Ball'.*
Miniature radio valve (two-volt, multi-electrode type)
Stephen Philip de Laszlo, *High Vacuum Valve Co, Holborn, London,* UK. *Called the 'Hivac Valve'.*
Paperback book
Sir Allen Lane *of London, whose Penguin books could be bought for the same price as ten cigarettes.*
Quick-setting ink (printing)
Dr D Askew, *chief research chemist of Coates Brothers Inks Ltd of St Mary Cray, Kent,* UK. *Their 'liquid phase separation' considerably speeded up the drying process.*
Radiolocation of aircraft (reliable system)
simultaneously by Robert Alexander Watson Watt, *Scottish physicist, Superintendent of the Radio Department of the National Physical Laboratory, Teddington,* UK, *and* Dr Rudolph Kühnold *of Kiel, Germany, and the* Société Française Radioélectrique, *Le Havre, France.*
Richter Magnitude Scale (earthquakes)
Charles Francis Richter, *seismologist, California Institute of Technology,* USA.
Semi-acoustic electric guitar
Lloyd Loar *for the Orville Gibson Guitar Co., Kalamazoo, Michigan,* USA. *It was called the 'Vivi-Tone'.*
Solar house heating
M Hottinger, *Zurich Institute of Technology, Switzerland.*
Strain seismograph
Hugo Benoiff, *Seismological*

Laboratory, Carnegie Institution of Washington, Pasadena, USA.
Tape-recorder (commercially available)
AEG GnbH, Germany. Called the 'Magnetophone'.
US Patent No. 2,000,000
Joseph Ledwinka *for the Edward G Budd Manufacturing Co., Philadelphia, Pennsylvania: safer construction for pneumatic railcar wheels.*
VHF electronic television
Marconi-EMI, UK, *using the 'Emitron' electronic camera.*

'Emitron' electronic television camera

Vinyl polymers
B F Goodrich Co., *Akron, Ohio,* USA, *following some nine years extensive research. By 1936 polymers became available on a large scale.*

1936
Australopithecus
Robert Broom, *S. African morphologist at Sterkfontein, South Africa.*

Boulder Dam
Colorado River, USA.
Electric iron (thermostat-control)
Messrs Morphy *and* Richards, UK.
Electro-mechanical calculator
Konrad Zuse, *Germany. By 1941 he was successfully operating his third prototype, Z3, a program-controlled electro-mechanical calculator.*
Intertype Fotosetter
C E Scheffer, *Swiss engineer, whose prototype was taken up by the* US *Intertype Corp. who then took a further ten years to perfect it under the direction of* H R Freund.
Jet-engine
simultaneously by Frank Whittle, RAF *officer-engineer at Cambridge University,* UK *and* Dr Hans von Ohain *in Germany. Whittle's theoretical work had begun some ten years before.*
Plastic lenses
Arthur Kingston, UK *photographer and inventor, who chanced upon a discarded test sample of* ICI *methyl methacrylate polymer (later Perspex) on sale in the Caledonian Market, London. From further research, Kingston helped to form Combined Optical Industries Ltd to manufacture plastic spectacle lenses, camera lenses, magnifying glasses, etc.*
Plexiglass contact lenses
I G Farben, *Germany.*
'Polaroid' (polarised plastic material)
Edwin Herbert Land, *physics undergraduate, Harvard University,* USA.
Practical helicopter (contra-rotating rotors)
Professor Heinrich Focke, *of Focke-Wulf, Germany – the Fa-61.*
Private automobile (diesel-engined)
Mercedes-Benz, *Stuttgart, Germany – the 260D.*
Spitfire (fighter-aeroplane)
Reginald Joseph Mitchell, *Chief Designer, Supermarine Aviation Ltd, Southampton,* UK *– the Supermarine Type 300. By 1946, over 22,000 Spitfires had been built and seen active service.*
Vitamin B6 (pyridoxine)
T W Birch *and* Albert von Nagyrapolt Szent-Györgyi, *Hungarian biochemist.*
Private automobile (diesel-Videophone service
The German Post Office, between Berlin, Nuremburg and Leipzig.
Volkswagen ('people's car')
Dr Ferdinand Porsche *of Stuttgart, Germany, at the behest of the Nazi Party. Factory built at Wolfsburg in 1938 for the 'KdF-Wagen', later the 'VW'.*

1937
Chairlift
Union Pacific Railroad engineers *in Chicago for the Dollar Mountain at Sun Valley, Idaho,* USA, *a resort famous for its ski-runs. It was 720 metres in length with a difference in level of 216 metres. Six other chairlifts were built for Dollar Mountain in quick succession. The first European chairlift was constructed in 1939 by* Nevrly, *an engineer at Pusterny, Czechoslovakia.*
Citric acid cycle
Hans Adolf Krebs, UK *biochemist and lecturer at Sheffield University.*
Drive-in-bank
Los Angeles, California, USA.
Electro-convulsive therapy
Doctors Ugo Cerletti *and* Lucio Bini *as a treatment for schizophrenia.*
Mirror reflex ciné camera
August Arnold *and* Robert Richter, *Munich, Germany. It was called the 'Arriflex 35'.*
Niacin
Conrad Arnold Elvehjem, *biochemist, University of Wisconsin,* USA, *and co-workers.*
Oil well (offshore)
Superior Oil Co. *and* Pure Oil Co., *in*

the Gulf of Mexico.
Photosensitive glass
R H Dalton *of the Corning Glass Works*, USA. *He found that if glass is exposed to ultra-violet radiation while it is still cold, the colour it eventually acquires is deeper.*
Radio telescope
Grote Reber, *of Illinois*, USA. *This was a dish, measuring 31 ft in diameter, for sending out and receiving back radio waves from outer space.*
Rayon yachting sail
used on the 'Ranger', UK.
Supermarket trolley (shopping cart)
Sylvan N Goodman *for the 'Humpty Dumpty' Supermarket, Oklahoma City*, USA.
Technetium
Emilio Gino Segré *and* C Perrier, *physicists, Palermo, Italy.*
Ticonal G (magnet type)
Philips Research Laboratories, *Eindhoven, Netherlands.*
Tinlet paintpot (model-making)
Douglas S Barton *of the Humber Oil Co., Hull,* UK – *to contain his Humbrol Art Oil Enamel paints.*
Titanium and zirconium
W J Kroll, *Luxembourg metallurgist, after seven years' research in his own private laboratory. The first manufacturing plant using the Kroll process, as developed by* F S Wartman *(*US *Bureau of Mines), was in operation at Boulder City, Nevada,* USA *by 1944.*

1938
Airliner (pressurised passenger cabin)
Boeing 307 Stratoliner, which went into service in 1940 for Transcontinental Airways.
Ballpoint pen
Ladislao J *and* Georg Biro, *brothers of Budapest, Hungary. Ladislao was a journalist and painter, whilst Georg was a chemist. On the outbreak of war,* the Biro brothers emigrated to Argentina where in 1943 with the help of Henry Martin, UK financier, they formed a company to perfect and produce the pen. It was at this stage that the familiar tube of ink and ball-bearing point were introduced. After the war, the Biro brothers' invention sold in its millions as the 'Eversharp' in the USA and ultimately as the 'Bic' in Europe.
Bren gun
the ZB26 as developed at Brno, Czechoslovakia.
Cortisone
simultaneously by Edward C Kendall *with* Dr Philip S Hench *of the Mayo clinic in Rochester, Minnesota,* USA, *and* Tadeus Reichstein *of the Technische Hochschule in Zurich, Switzerland, after almost twenty years' concurrent research into a drug which would alleviate Addison's disease or rheumatoid arthritis.*
Domestic steam iron (thermostatic control)
Edmund Schreyer, *Ridgefield, Connecticut*, USA.
Ejection seat
Junkers Aircraft Factory, *Germany, although three other German aircraft works were also experimenting with the idea. The first aircraft to be fitted was a Junkers 88 in 1939. By 1940 more than 200 test ejections had been made using a compressed air gun to fire seats up rails from ground test rigs, some 70 of them with human subjects. The Luftwaffe began to use ejection seats in 1941 and by 1943 German aircrews were using the seat in battle emergencies. The* UK *Martin-Baker ejector seat was not tested from the air until 1946.*
Fluorescent lighting
Dr Arthur H Compton, *Professor of Physics, Chicago University and consultant to the* US *General Electric Co., with* Dr George Inman, *head of GEC's research team. Work began on the project in 1934. Unknown to GEC,*

1938

a similar lamp had been invented in *1926, by three Germans –* Meyer, Spanner *and* Germer *of the Rectron Co. – but they had not developed their invention.*

Flying-boat-launched seaplane
Short Brothers *of Rochester,* UK *– the 'Composite', with the 'Mercury' taking off from the 'Maia', in flight.*

Folacin
P L Day.

High definition colour television (mechanical)
John Logie Baird, *Long Acre, London, who demonstrated his 'Telechrome' in a closed laboratory-studio.*

Instant coffee
Nestlé *of Vevey, Switzerland, following eight years of research after a suggestion made by the Brazilian Institute of Coffee. It was trade-named 'Nescafé'.*

Language laboratory (wire-recorder based)
First tested at Middlebury College, Vermont, USA. *Based on the 'Sound Mirror' process developed by* Marvin Camras *at the Armor Research Foundation, Illinois Institute of Technology. Marketed by Brush Development Corporation. Camras's 'record and erase' process was developed three years later.*

Nuclear fission
Otto Hahn *and* Fritz Strassman, *Kaiser Wilhelm Institute of Chemistry, Berlin.*

Nylon
Dr Wallace Carothers, *leading a team of organic chemists at the Eastern Laboratory of Du Pont de Nemours, Repauno, Delaware,* USA. *$27,000,000 was invested by Du Pont in nylon research over thirteen years before their factory at Seaford, Delaware could mass-produce it. Although at first twice as expensive as silk, it was also twice as durable.*

Nylon toothbrush
Dr West's *'Miracle Tuft Toothbrush',* USA.

Perlon (polyamide)
P Schalk, *Germany.*

PTFE (polytetrafluorethylene)
Roy J Plunkett, *research chemist at the Kinetic Chemical Co.,* US.

Pressure-cooker (self-sealing, interlocking pan and lid)
Alfred Vischer, *Chicago draughtsman.*

Sulphonamide drugs
research team at May & Baker's laboratories *in Dagenham, Essex,* UK. *After considerable experimentation, they produced M&B 693 or 'Sulphapyridine' to arrest the growth of pneumococcus, the organism responsible for pneumonia.*

Television camera for outside broadcasts
Sidney Rodda *and* Hans Lubszynsky *of Marconi-EMI for the BBC,* UK. *Called the 'Super Emitron'.*

Wein Bridge Oscillator (stabilized version)
William Hewlett, *electrical engineer,* and David Packard, *formerly of the Vacuum Engineering Department, General Electric Co., Schenectady, New York,* USA. *With an order for 8 of their '200 A Audio Oscillators' from Walt Disney, the Hewlett-Packard Co. was formed in Palo Alto, California.*

Xerographic photo-copying machine (prototype)
Chester Carlson, US *physicist. The first-ever copy read '10-22-38 Astoria'. It was perfected from 1944 by* Roland M Schaffert, *research physicist for the Battelle Memorial Institute, Columbus, Ohio,* USA.

1939

Betatron
Donald W Kerst, *Assistant Professor of Physics, University of Illinois,* USA.

Binary calculator
John Atanasoff *and* George R Stibitz, *mathematicians at Bell Telephone Laboratories,* USA.

DDT (insecticide)
Paul Hermann Müller, *German chemist at J R Geigy, Basle, Switzerland. Trade-named 'Gesarol', it was to be used in large quantities to kill fleas and body lice amongst troops and refugees.*

'Electric aura'
Semyon Kirlian, *Soviet electrician, while repairing an electrotherapeutical instrument. For the next 30 years, Kirlian battled to perfect his apparatus, capable of 'photographing' the aura around the human hand.*

Jet-engined aircraft
Heinkel He 178, with a powerplant built to Dr Hans von Ohain's *design. By 1941* Frank Whittle's *W I engine, as produced by Power Jets (UK), was installed in the successful Gloster E28/29. By 1944 the* RAF *had formed a squadron of jet aircraft and the Germans had one engine, the Junkers 004, in production.*

Gloster E28/29 aircraft

Microfilm camera (mass card-filing system)
Elgin G Fassel *of Milwaukee, Wisconsin. It was marketed by the Microstat Corporation of America.*

Nylon 'Supra' spectacles
Neville Chappell, *manufacturing optician, Amersham, Buckinghamshire,* UK *– the first spectacles to have a continuous top rim and bridge, and subsequently to use nylon.*

Oil (commercial quantities – UK)
British Petroleum Co. Ltd *at Eakring, Nottinghamshire.*

Polyurethane paint
Otto Bayer, *Germany.*

Practical helicopter (single main and small tail rotors)
Igor Sikorsky, *Soviet émigré at the Vought-Sikorsky Co. in Bridgeport, Connecticut,* USA *– the V S-300.*

Speech-simulating machine
Bell Telephone Laboratories, Holmdel, New Jersey, USA. *Exhibited at the New York and San Francisco World Fairs.*

Vaccine 17-D (yellow fever)
Max Theiler, *South African bacteriologist, International Health Division, Rockefeller Institute of Medical Research, New York,* USA.

1940

Amphibious Volkswagen
Dr Ferdinand Porsche, *Stuttgart, Germany Type 128, called the 'Schwimmwagen'.*

Automatic gearbox
General Motors, *Detroit,* USA.

Cavity magnetron (centimetre radar valve)
John Turton Randall, *Professor of Physics, King's College, University of London, with* James Sayers *and a team at the Birmingham University Physics Laboratory.*

Deuterium (heavy-hydrogen isotope)
Hans von Halban *and* Lew Kowarski, *French physicists. It was called 'heavy water'.*

Extra-sensory perception
Joseph Banks Rhine, *psychologist and professor at Duke University,* USA.

'Flying wing' aircraft (tailless)
US Northrop Aviation's *N-1 M.*

Helicopter (production model)
Focke-Aghelis-223, with a passenger capacity of six persons.

Jeep
Karl K Pabst, *consulting engineer of the Bantam Car Co., Butler, Pennsylvania, for the* US *Army. Powered by a 45 bhp, four-cylinder Continental engine, the first batch of 70 went into service the following year. 649,000 jeeps were produced during the war.*

'Mae-West' lifejacket
RAF *Mark I inflatable lifejacket.*

Neptunium
Edwin Mattison McMillan, *physical chemist, and* Philip Hauge Abelson, *Radiation Laboratory, University of California, Berkeley,* USA.

Orgone accumulator
Wilhelm Reich *of Long Island, New York,* USA – *a mini-greenhouse capable of capturing 'Orgone' or basic life energy and increasing the vitality of anyone sitting in it.*

Penicillin (stabilised and quantity-produced)
Dr Howard Walter Florey, *Australian pathologist, and* Dr Ernest Boris Chain, *German-refugee chemist, working at Oxford University,* UK: *produced in quantity by* US *drug firms.*

Plutonium
Glenn Theodore Seaborg *and* Edwin Mattison McMillan, *nuclear scientists, Radiation Laboratory, University of California,* USA.

Radar (radio detection and ranging)
word coined by Commander S M Tucker, US *Navy, to describe innovations made since 1936 by* Robert M Page, US *Naval laboratory.*

Radio-controlled tank
Marcel Pommelet, *French engineer.*

Rhesus blood group system
Karl Landsteiner *and* Alexander Solomon Wiener, *haematologists at the Rockefeller Institute of Medical Research, New York,* USA.

Rocket-powered glide bomb (radio-controlled)
Dr Herbert Wagner, *Germany – the HS293.*

Vitamin H (biotin)
Vincent du Vigneaud, *biochemist, professor at Cornell University Medical College,* USA.

1941

Aerially-launched fighter-planes
Soviet Air Force's Zveno 'parasite'. In experiments, three Polikarpov 1-5 biplanes were wing-launched from an in-flight Tupolev TB-3 heavy bomber.

Aerosol insect spray (carbon-dioxide propulsion)
L D Goodhue *and* W N Sullivan, US.

Anti-tank grenade projector
Watts *and* Jeffries, UK. *Based on the work of* Lieutenant Colonel Blacker, *it was called* PIAT *(projector, infantry anti-tank).*

'Bat bombs'
Louis F Fieser *and* nineteen university professors *on the* US *National Defense Research Committee. Thousands of artificially hibernated bats, an incendiary bomb attached to each one, would be released from aircraft bomb bays. Awoken by the warm air, they would fly into buildings where the incendiaries would ignite. Planned for Japanese cities, they were never used.*

Battery (silver-zinc)
H André, *France.*

Cardiac Catheter
Dick W Richards US, *and* André F Cournand, *France at Bellevue Hospital, New York. By developing a crude rubber tube originally used in 1929 by* Dr Werner Forssmann, *Germany, for investigating diseased hearts.*

Colour TV service (experimental)
Dr Peter Goldmark *of Columbia Broadcasting Service, New York.*

Electronic autopilot
Minneapolis-Honeywell Co., USA – *called the C-1.*

Electronic flight simulator
Dynatron Radio Ltd, *Hampton Court, Surrey,* UK.

Giant windmill (electricity-generating)
a 1250 Kilowatt windmill was erected at 'Grandpa's Knob' in the central Vermont mountains, USA. *It had twin blades, 175 ft in diameter. The experiment ended when a blade cracked*

after three years' service.

Locomotive (gas-turbine-engined)
Switzerland.

MCPA (herbicide)
Messrs Slade, Templeman *and* Sexton *at Jealott's Hill Research Station,* ICI, UK.

Microwave radar set
US *Radiation Laboratory.*

Particleboard (commercial)
Heseke *of Torfitwerker* AG, *Bremen, Germany – from wood chips and thermosetting resins. Worldwide manufacture did not begin until* 1946.

Portable two-way FM radio
Motorola, USA. *Called the 'Handie/Walkie-Talkie'.*

Shell-moulding ('C' process)
Johannes Croning, *proprietor of a Hamburg foundry, using a thin mould of sand held together by phenolic resins. By* 1948 *Ford of Detroit had a pilot line shell-moulding* 1000 *exhaust valves per day.*

Television advertisement
WBNT – *New York, for the Bulova Watch Co.*

Terylene
John R Whinfield, *Cambridge chemistry graduate, and his assistant* James T Dickson *at the research laboratories of the Calico Printers' Association, Accrington, Lancashire,* UK. *It was marketed and mass-produced by* ICI *in the* UK, *and under licence by Du Pont de Nemours,* US *first as 'Fiber V' then as 'Dacron'.*

2,4-D (herbicide)
Messrs Nutman, Thornton *and* Quastel *at Rothamstead Experimental Station, Harpenden, Hertfordshire,* UK.

1942
Atomic pile reactor
Enrico Fermi, *Italian refugee physicist,* and a team *working at Stagg Field, University of Chicago,* USA *(known as the Manhattan Project).*

Bazooka gun
Lieutenant-Colonel Leslie A Skinner (US Army) *to fire 2 ft long ground rockets from a steel tube; the charge was set off by an electric battery. It was officially known as 'Launcher, Rocket,* AT, M-1 *'.*

DUKW (amphibian)
Sparkman *and* Stephens, *together with* General Motors Corporation Truck and Coach Division, US.

GR-S (government rubber styrene)
US Government, *following the Japanese over-running of Malaya and then the Dutch East Indies to cut off main sources of rubber from the* USA. *As a substitute, this synthetic rubber was manufactured from butadiene and styrene.*

Napalm (incendiary)
Louis F Fieser *at the* US *National Defense Research Committee's laboratories. It was later used to saturate Japanese cities in the closing weeks of the war.*

Supersonic long-range war rocket
General Walter Dornberger, Eugen Sänger, Wernher von Braun, Hermann Oberth *and* 2000 *research workers at the Rocket Research Station, Peenemünde, on the Nazi Baltic Coast. Over* 35,000 *such V 2 rockets were launched on London and Antwerp.*

1943
Aqualung
Commandant Jacques-Yves Cousteau, *French Navy, with* Emile Gagnan, *control-valve engineer.*

Bailey bridge (prefabricated steel bridge for rapid temporary use)
Donald Coleman Bailey, UK *engineer.*

Bouncing bomb
Barnes Wallis, *Vickers Aviation Ltd, Weybridge, Surrey,* UK. *Four years in development, and inspired by the memory that Admiral Horatio Nelson*

1943

had had an idea of aiming his cannons at the sea, so as to bounce up the shot and hit the French warships, Wallis's bombs successfully destroyed the Mohne, Ader and Sorpe dams on the Ruhr.

LSD (drug)
Dr Arthur Stoll, research director of the Swiss Sandoz pharmaceutical company, whilst synthesising the first natural ergot alkaloid. His assistant, Dr Albert Hofman, chemist, was unknowingly the first to undergo the hallucinatory effects of this drug. Although originally manufactured for use by qualified professionals investigating mental disease, LSD also fell into the hands of drug addicts and was soon known as a 'killer'.

Programmable electronic computer
Professor Max H A Newman and T H Flowers at Bletchley Park, Buckinghamshire, UK. Called 'Colossus', it used 1500 valves and was operated to decipher the output of the German coding machine, Enigma.

Sailing trimaran
Victor Tchechett, Soviet émigré to the USA.

Streptomycin (antibiotic drug)
Dr Selman Abraham Waksman, Russo-Jewish-born, US-trained bacteriologist and biochemist, with his students at Rutgers University, New Jersey, USA. By 1944, Merck & Co. Laboratories had developed a deep vat fermentation process for its mass-production.

Teflon
Du Pont de Nemours, US.

1944

Air-to-air missile
Dr Herbert Wagner, Germany. The HS 298 was designed to be carried by either the Do 217 or FW90 aircraft.

Americium
Glenn T Seaborg and Albert Ghiorso, University of California, Berkeley, USA.

ASCC Mark I (automatic sequence-controlled calculator)
Professor Howard H Aiken of Harvard University, US in association with International Business Machines (IBM) engineers at Endicott, New York.

Curium
Glenn T Seaborg and Albert Ghiorso, University of California, Berkeley, USA.

Decca navigator-receiver QMI
used by the Royal Navy for mine clearance.

Forward-swept-wing aircraft (jet-engined)
Junkers Ju 287, built at Dessau, Germany.

Ground-to-air missile
Dr Wurster, Germany – 'Enzian'.

Marine wave energy absorber and diverter
Robert Lochner, Royal Naval officer. Called 'Bombardon', it was used during the D-Day landings.

Nerve gases
Nazi chemists, but never used during conflict.

Pyrex telescope lens (diameter 200 inches, weight 15 tons)
Corning Glass Works, to the design of George Ellery Hale, US astronomer, for the Mount Palomar Observatory. Casting took 7 hours, cooling took 10 months, and polishing – using 30 tons of grinding paste – took 11 years. Although the lens was in position by 1944, the telescope was not in operation until 1949.

1945

Artificial kidney
Dr Willem J Kolff, Dutch surgeon. His early artificial kidneys were made of cellophane.

Atomic bomb
Professors J R Oppenheimer, Arthur H Compton, Enrico Fermi and Léo Szilard at the Los Alamos Laboratory near Santa Fé, New Mexico. Following a test explosion at

Alamogordo, two bombs were dropped on Japan (Hiroshima and Nagasaki) with 130,000 victims.

Cephalosporium Acremonium (isolated)
G Brotzu, *Cagliari Institute of Hygiene, Italy.*

Contact lens (fenestrated solution-less)
Norman Bier, UK.

Flying ram aircraft
Northrop XP-79, US.

Microwave oven
Percy LeBaron Spencer; *manufactured by the Raytheon Co., Waltham, Massachusetts in 1947.*

Nylon sail.

Programming language (for electro-mechanical calculator)
Konrad Zuse, *Germany. He lost all but his Z4 machine in air-raids. He called his language 'Plankalkül'.*

Surface-to-air missile
US *Western Electric Company's Nike-Ajax.*

Torsion bar suspension
Douglas Motorcycle Co., *Bristol,* UK, *for their 350 cc Flat Twin model.*

Tupperware boxes (plastic)
Earl W Tupper, *former chemist with Du Pont de Nemours,* USA.

Vinyl floor covering.

Zoom lens
Dr Frank Back, *Zoomar Corporation, New York State. It was first used to cover the baseball world series as broadcast on* US TV *in 1947.*

1946

Artificially induced rainfall
Vincent J Schaefer *and* Irvin Langmuir *of General Electric Co., Schenectady, New York. Carbonic snow was dispersed into a cloud over Greylock, Massachusetts,* USA.

Artificial ski ramps
Major Tony Moore, *Staff Officer for Physical Training with the British Army at Klagenfurt, to 'break in' trainee soldiers to the disciplines of alpine and cross-country skiing.*

Bikini swimsuit
Louis Reard, *French couturier. It was first modelled by a dancer called Micheline Bernardi at a Paris fashion show, four days after the* USA *had detonated an atomic bomb at Bikini Atoll in the Pacific.*

Commercial radiotelephone service
between the public telephone system and motor vehicles, USA.

Electric window lift (automobile)
Cadillac and Lincoln, US, *and Daimler,* UK.

ENIAC (electronic numerical integrator and calculator)
Dr John W Mauchly *and* J Presper Eckert *of the University of Pennsylvania,* USA. *With 18,000 vacuum tubes, this was the first* US *vacuum-tube computer.*

Espresso coffee machine
Gaggia, *Milan, Italy.*

Kell blood group system
Race, R R Coombs *and* A E Mourant.

Lutheran blood group system
Race *and* Callender.

Photo-typesetting machine
René Higonnet *and* Louis Moyroud, *French telephone engineers, following two years' work in their own homes Backed first by the* US *Lithomat Corp. then by the Graphic Arts Research Foundation, development of this machine continued throughout the 1950s and it was ultimately marketed as the Photon in the* USA *and the Lumitype in Europe.*

Radial-ply tyre (steel-reinforced)
Pierre-Marcel Bourdon *and* Marius Mignol, *Michelin et Cie, Clermont-Ferrand, France. Called the 'Michelin X', it was developed from the 'Pilote' and 'Metallic' tyres produced by this company 10 years earlier.*

Vinylite disc record
RCA-Victor, USA, *to replace records made of shellac.*

1947

Coenzyme A
Fritz Albert Lipmann, *Head of the Biochemical Research Department, Massachusetts General Hospital, Boston*, USA.

Flying-boat fighter (turbojet-engined)
Saunders Roe SR A/1, *Cowes, Isle of Wight*, UK.

Flying-wing jet bomber
Northrop YB-49, *powered by eight Allison* J-35 *turbojets*, US.

Gleep (graphite low energy experimental atomic pile)
Harwell, UK.

Holography (3-D)
Dennis Gabor, *Hungarian-born Professor of Electrical Engineering at Imperial College, London. He made his first experiments with mercury lamps, giving his system the name holography after the Greek 'holos' (whole).*

'Intelligent machinery'
Alan M Turing, *logician at Cambridge University*, UK. *His prophetic article remained unpublished for 30 years.*

Land-Rover
The Wilks brothers Spencer *and* Maurice, *had the idea of producing a go-anywhere vehicle at the latter's farm in Anglesey, Wales*, UK. *Since then, Land-Rovers have travelled across the globe.*

Motorboat (gas-turbine-engine)
Metropolitan-Vickers Electrical Co. Ltd, *who fitted a 2500 hp gas turbine into the British Admiralty triple-screw* MGB 2009.

Pernicious anaemia (Vitamin B12 as cure for)
Dr Karl August Folkers, *chemist working for Merck & Co., Rahway, New Jersey*, USA, *with assistance from* Dr Mary Shorb *of Maryland University.*

Pilotless transatlantic aircraft
Douglas Aircraft, US. *A* DC-4 *Skymaster flew some 2400 miles.*

Pion
Cecil Frank Powell, *Professor of Physics at Bristol University*, UK, *in conjunction with* Giuseppe Ochiatini, *Italy.*

Polaroid Land camera
Dr Edwin H Land, *who had originally formed his Polaroid Corporation in Cambridge, Massachusetts to manufacture polaroid sunglasses. The automatic developing and printing process was carried out in the camera itself. The polaroid colour photo, trade-named 'Polacolor', was introduced in 1963 – a direct positive being produced in one minute.*

Powerboat (converted to turbojet-propulsion)
Peter du Cane, *naval architect, Portsmouth*, UK *and* Major Frank Halford, *engineer, Hatfield*, UK, *who modified Sir Malcolm Campbell's 'Bluebird K4' hydroplane to take a de Havilland Goblin II jet engine.*

Promethium
J Marinsky *et al.*

Radiocarbon dating
Dr Willard Frank Libby, *Professor of Chemistry, Institute for Nuclear Studies, University of Chicago*, USA.

Supersonic aircraft
the Bell XS-1 rocket-plane, *'Glamorous' Glennis', launched from a Boeing B-29 Superfortress in flight and piloted by Captain 'Chuck' Yeager over Edwards Air Force Base, California. Its top speed was Mach 1.015 (670 mph).*

Synthesised Vitamin A
O Isler.

Tractor (power adjustments of rear wheel tread)
Allis Chalmers, US.

Tubeless tyres
B F Goodrich Co. *of Akron, Ohio*, USA.

Tungsten carbide dental drill.

1948

Atomic clock
Dr Willard Frank Libby, *Professor of Chemistry, Institute for Nuclear Studies, University of Chicago*, USA. *First model built at the National Bureau of Standards, Washington* DC.

Cybernetics
Norbert Wiener, US *mathematical logician, Massachusetts Institute of Technology.*

McDonald hamburgers
Maurice *and* Richard McDonald, *brothers in Pasadena, California*, USA.

Microgroove record (long-playing)
Dr Peter Goldmark *and* a fifteen-strong research team *at the Columbia Broadcasting System*, USA. *The 33⅓ rpm speed was soon to oust the 78 rpm monopoly. Soon after, the Radio Corporation of America introduced a 45 rpm microgroove disc for 'pop' as opposed to 'classical' music.*

Passenger elevators (electronically controlled)
Otis Co. *at Universal Pictures Building, New York.*

Polyester-resin boatbuilding
Dr Irving E Muskrat *of Marco Resins Inc., New Jersey*, USA.

Rear-window heating (embedded tungsten wires)
Rolls-Royce Cars Ltd, UK.

Relascope (tree-measuring instrument)
Walter Bitterlich, *forestry officer of Salzburg, Austria, for angle-count sampling.*

'Scrabble' (board game)
James Brunot, *Connecticut*, USA.

Solid electric guitar (production model)
'Leo' (Clarence) Fender, 'Doc' Kauffman *and* George Fullerton, *of California. It was called the 'Fender Broadcaster', later renamed the 'Telecaster'.*

Soluble aspirin
Reckitt & Sons Ltd, *Hull*, UK – *called 'Disprin'.*

Transistor (point contact)
Dr William Shockley, John Bardeen *and* Walter H Brattain *at the Bell Telephone Laboratories*, USA.

Turboprop airliner
Rex K Pearson *and* George Edwards *at Vickers Aviation Ltd, Weybridge, Surrey*, UK. *The 'Vickers Viscount' was powered by four Rolls-Royce Dart turboprop engines.*

Universal digital stored program computer
Professors Sir Frederick Williams *and* T Kilburn *at Manchester University*, UK. *Called the 'Manchester Mark I', and using electrostatic tubes, it was manufactured by Ferranti Ltd in 1951.*

'Velcro' (fastening device)
Georges de Mestral, *Swiss engineer.*

1949

Aircraft (delta-wing)
the Avro 707, UK.

Berkelium
Glenn Theodore Seaborg *with* Stanley Gerard Thompson, *nuclear scientists at the Lawrence Radiation Laboratory, University of California, Berkeley*, USA.

Cable television (experimental)
Oregon, USA.

Colour television tube
David Sarnoff *and* a research team *at the Radio Corporation of America – called the 'Shadowmask'.*

EDSAC (electronic delay storage automatic calculator)
researchers at *Cambridge University*, UK.

EDVAC (electronic discrete variable automatic computer)
John von Neumann and a team *at Pennsylvania*, USA.

High-altitude liquid-fuel rocket
Viking, *fired to an altitude of 50 miles in 54.5 seconds, over White Sands, New Mexico.*

Jet-powered airliner
design team led by R E Bishop *of the*

de Havilland Aircraft Company, Hatfield, UK. *The DH 106 Comet was powered by the de Havilland Ghost 50 engine.*
Key starting (automobiles)
the Chrysler Corp., *Detroit, Michigan,* USA.
Magnetisable ferrite cores (for data storage)
Jay Wright Forrester *of the Massachusetts Institute of Technology, Boston, for the 'Whirlwind' computer.*
'Super amp' (rock music amplifier)
Leo Fender *and* Donald Randall, US.

1950
Californium
Glenn T Seaborg *and* Albert Ghiorso, *nuclear scientists at the Lawrence Radiation Laboratory, University of California, Berkeley.* USA.
Chlorpromazine (tranquilliser drug)
Paul Charpentier, *French chemist working at the Rhône-Poulenc laboratories of the Specia drug firm. It took him only three months to synthesise a new phenothiazine derivative, which was to have remarkable effects on schizophrenic patients.*
Credit card (for general purchase)
Ralph Scheider *for the first 200 members of the New York Diners' Club.*
Duffy blood group system
James S Cutbush, P L Mollison *and* D Parkin, *chemists,* US.
'Elmer and Elsie' (robots)
Dr Gray Walter *of the Burden Neurological Institute, Bristol,* UK. *The robots, made of photo-electric cells, batteries, transistors, relays and electric motors, were capable of making simple decisions.*
Ferroxdure (ferrite)
Philips Research Laboratories,

Eindhoven, Netherlands.
Gas-turbine car
Rover Ltd, UK – *the prototype two-seater Jet 1.*
Oceanographic research ship
'Calypso', commanded by Jacques Yves Cousteau *of the French Navy's Undersea Research Group.*
Radio-immunoassay
Rosalyn Sussman Yalow, *Veterans Administration Hospital, Bronx, New York.*
Schottel-Rudderpropeller
Joseph Becker, *Rhineland boatbuilder.*
Xerographic-copying machine (production 'office model')
Haloid Co. *of New York – called the 'Xerox 914'.*

1951
Cinerama
Fred Waller, US, *making use of three projectors and a wide curved screen to recreate a stereoscopic motion picture experience. Initial tests had been going on since 1936 at a former indoor tennis court near Oyster Bay, Long Island, New York. In 1952 'This Is Cinerama' opened in a Broadway cinema and ran for 2½ years. Its seven sound tracks were developed by Hazard Reeves.*
Experimental breeder reactor No. 1
Walter H Zinn, *one of Enrico Fermi's wartime assistants at the National Reactor Testing Station in central Idaho,* USA. *This was the first nuclear reactor to produce a significant amount of electric power.*
Helicopter (gas-turbine-engined)
SO 1120 *'Ariel III', France.*
Isoniazid (drug)
simultaneously by a team of chemists at the Hoffmann-LaRoche *pharmaceutical company, New Jersey, and by a* team of scientists at E R Squibb & Son *laboratories in New Brunswick, both* USA. *Isoniazid was*

first used with streptomycin on tuberculosis patients by Doctors Selikoff and Robitzek at Sea View Hospital, Staten Island, New York, the world's largest TB hospital. Having successfully tested it on themselves, the doctors administered it to their worst 92 patients, nearly all of whom showed rapid signs of recovery.

Merchant ship (converted to gas-turbine engines)
'Auris' by John Lamb, Head of Research and Development at Anglo-Saxon Petroleum Ltd, UK.

Mobile genetic elements
Barbara McClintock, cytogeneticist, Cold Spring Harbor Laboratory, New York, USA, while studying the maize plant.

Motor-cruiser (glass-fibre)
M Rosenblatt & Son, US. The 'Dreadnought 23' was built using the Marco vacuum process.

Oil (Persian Gulf)
Aramco, who discovered the great Safaniya field.

Oral contraceptive pill
Gregory Goodwin Pincus and Min Chuch Chang, biologists of the Worcester Foundation for Experimental Research, Shrewsbury, Massachusetts, in collaboration with Dr John Rock, obstetrician-gynaecologist of the Brookline Reproductive Clinic, Boston, and again in collaboration with Dr Carl Djerassi of the Syntax Corporation, Mexico, who discovered the progestogenic agent, 19-Norsteroids. The US Food and Drug Administration approved 'the pill' for public use in 1961, after extensive trials in Puerto Rico and Haiti.

Power-steered automobile
Chrysler, USA.

Precision-base electric guitar
Leo Fender and 'Doc' Kauffmann, California, USA.

Transistor (junction)
Dr William Shockley, R L Wallace and Morgan Sparks at the Bell

Transistor

Telegraph Laboratories, USA. The original point-contact transistor was found to be noisy and unable to control high amounts of power, and had limited applicability. It was therefore superseded by the junction-transistor.

1952

Artificial heart valve
Dr Charles A Hufnagel at Georgetown University Hospital, Washington DC, USA.

Automated automobile manufacture (partial)
the Ford Co. at Cleveland, Ohio, USA.

Bubble chamber
Donald Arthur Glaser, physicist, at the University of Michigan, USA.

Cosmotron (3 billion-volt proton synchroton)
Brookhaven National Laboratory, Long Island, New York.

Electric coffee pot (fully automatic)
Bill Russell and Peter Hobbs of Croydon, UK.

Electron (precise magnetic moment of)
Polykarp Kusch, Professor of Physics at Columbia University, USA.

Electronic calculator (germanium diodes as electrical components)
Compagnie Bull, Paris – called the 'Gamma 3'.

FM car radio
Blaupunkt, *Germany, – with push-button tuning.*

Hearing aid (transistorised)
Sonotone Corp. *of Elmsford, New York.*

Hydrofoil (turbojet-engined)
Frank Hanning-Lee, UK. *The 'White Hawk' was unsuccessful in trials on Lake Windermere.*

Hydrogen bomb
simultaneously by Edward Teller, US, *and* Igor Kurchatov, USSR. *The US test (6 megatons of TNT) was at Eniwetok Atoll, Pacific Ocean, and the Soviet test was in the Arctic in 1953.*

Integrated circuit (theoretical)
G W A Dummer, *Royal Radar Establishment, Malvern,* UK.

Pattern 5580 inflatable lifejacket
Dr E A Pask, RAF *Physiology Laboratory,* RAF *Farnborough,* UK, *after ten years' research and development.*

Powerboat (custom-built for turbojet power)
Peter du Cane *of Vospers Ltd,* Reid A Railton, *aerodynamicist, and* Major Frank Halford, *engineer. Powered by a de Havilland Ghost turbojet, 'Crusader' disintegrated at 240 mph on Loch Ness, Scotland, killing her pilot,* John Cobb.

Surface-to-surface missile
the 'Firestone Corporal', US.

Terylene yachting sail
Used on 'Sonda' of West Mersea, UK.

UNIVAC I (universal automatic computer)
Eckert and Mauchly Computer Co., *Philadelphia, as the first computer designed for mass production. Built by Remington Rand, it was delivered to the* US *Census Bureau.*

Videotape (experimental)
John Mullin *and* Wayne Johnson *at Bing Crosby Enterprises Laboratories, Beverly Hills, California,* USA.

'WD 40' (water-displacer and temporary rust preventative)
Norm Larson, *for the Rocket Chemical Co., San Diego, California,* USA. *It was first used on certain parts of the Atlas missile. The aerosol version of 'WD40' was available by 1955.*

1953

Automobile (glass-fibre bodywork)
Chevrolet Corvette, US.

'Cloudbuster'
Wilhelm Reich, US, *channelling his orgone energy through a bank of metal tubes to disperse hurricanes,* UFOs *and black clouds, and to make rain where there was drought – questionable.*

Collapsible polythene tube
Bradley Container Corp. *of Delaware,* USA, *for 'Sea and Ski', a skin-tanning lotion.*

Commercial passenger hydrofoil
Baron Hanns von Schertel, *Germany, of Supramar, the pre-war Schertel-Sachsenberg Research and Development Consortium. It first plied for service on Lake Maggiore between Switzerland and Italy.*

DNA (deoxyribonucleic acid)
Francis Harry Compton Crick, *molecular biologist,* UK, *and* James Dewey Watson, *biochemist,* US, *at the Medical Research Council Unit, Cavendish Laboratory, Cambridge,* UK.

Disc brakes (aircraft).

Einsteinium
Albert Ghiorso *et al., Heavy Ion Linear Accelerator, Lawrence Radiation Laboratory, University of California,* USA.

Fermium
Albert Ghiorso *and* Stanley Gerald Thompson, *Lawrence Radiation Laboratory, University of California,* USA.

Heart-lung machine
Dr John H Gibbon *at the Thomas Jefferson University Hospital, Philadelphia, Pennsylvania,* USA, *after some eighteen years' research.*

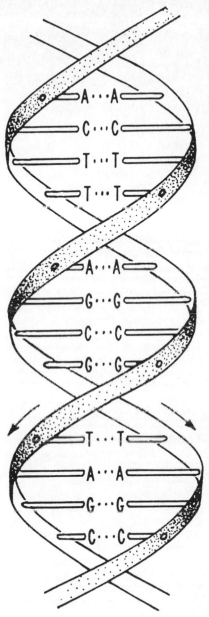

DNA molecule

Jet-prop helicopter
independently, Fairey Aviation Co.'s
Gyrodyne, UK, *and the* SO*1221* Djinn,
France.
Measles vaccine
Dr John Franklin Enders *and his
assistant*, Dr Thomas Peebles,
*paediatricians at the Children's
Hospital, Boston, Massachusetts*, USA.
*Peebles was able to isolate a measles
virus from the blood and throat
washings of a schoolboy.*
Nylon marine propeller
Ajax Marine Engines Ltd, UK.
Rapid computer print-out
Remington Rand, US, *for* UNIVAC. *It
was capable of 600 lines per minute.*
**Reserpine (anti-depressant
drug)**
Dr Nathan S Kline, *psychiatrist at
Rockland State Hospital,
Orangeburg, New York. Three years
later, he discovered iproniazid as a
second anti-depressant drug.*
Swing-wing aircraft
Grumman, XF 10 F *Jaguar jet-fighter*
US. *Only two prototypes were ever
built.*
Supercocotte (pressure cooker)
Lescure brothers (Frederic, Jean and
Henri), *for* SEB, *Boulogne, France.*

1954
**Atomic absorption
spectrophotometer**
Dr Alan Walsh, *Chemical Division of
the Commonwealth Scientific and
Industrial Research Organisation,
Melbourne, Australia.*
Ball-bearing handpiece (dental)
*enabling the drill to spin at 25,000
rpm.*
**Colour television (regular
broadcasts)**
the National Television System
Committee, USA.
**Computer memory magnetic
cores (commercial)**
IBM *at Poughkeepsie, US, for their 704
and 705 computers, developed in
conjunction with* Massachusetts

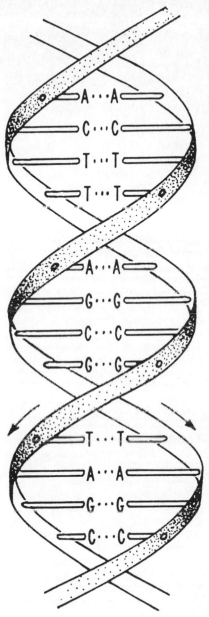

DNA molecule

1954

Institute of Technology's Lincoln Laboratory, *based on the pioneering work of* A Wang *and* F W Viehe.

Electronic music synthesiser
Dr Harry Olsen *of the Radio Corporation of America; re-designed and installed at both Columbia and Princetown Universities as the* RCA Mark II.

Hatchback estate car (side opening rear door)
Hillman Husky, UK.

High-speed switching surface barrier transistor
Philco, US.

Input/Output computer channel
Bob Evans *on* IBM 704.

Isotactic polypropylene
separately by Karl Ziegler, *German chemist, Director of the Max Planck Carbon Research Institute at Mülheim, and* Giulio Natta, *Italian chemist, at the Milan Institute of Technology. It was first marketed by Montecatini as 'Moplen' in* 1957.

Maser prototype
Charles Hard Townes, *physics professor, with* James P Gordon *and* Herbert J Zeiger *at Columbia University,* USA. *Also by* Aleksandre Mikhailovic Prochorov *with* Nikolai Gennadievic Bosov *at the Lebedev Physics Institute,* USSR. *The practical maser (three-level, solid state) was developed by* Nicholas Bloembergen *at Harvard University and built by Messrs Feher, Scovil and Seidel at Bell Telephone Laboratories,* USA.

Motor yacht (glass-fibre construction)
49ft 'Perpetua', moulded in 'Deborine' for Patrick David de Laszlo *at his own company, Halmatic Ltd of Portsmouth,* UK. *Given strength by de Laszlo's patent 'Top Hat' aluminium framework for her ribs and stringers. 'Perpetua' was fitted out with twin Gardner diesel engines.*

Non-stick pan
Marc Grégoire, *French research engineer at* ONERA, *by coating Téflon*

Patrick De Laszlo's 48ft glass-fibre 'Perpetua'

on to metal. The Téfan Co. was formed in 1956.

Passenger conveyor (permanent)
'Speedwalk' at Erie Station, Jersey City, USA.

Retractable ball-point pen (press-and-press-again)
Parker Pen Co., *Janesville, Wisconsin,* USA.

Rotary piston engine
Felix Wankel, *motor engineer from Lindau, Bavaria, after 30 years' research; production engine by* NSU Motorenwerke *in* 1957.

Solar battery (400 silicon cells)
G M Chapin, C S Fuller *and* G L Pearson *at Bell Telephone Laboratories,* USA.

Supersonic delta-wing seaplane
Convair's *jet-engined 'Sea Dart' for the* US *Navy.*

Transistor radio
Regency Electronics, *Indianapolis,* USA – *the 'TR-1'.*

Tubular-nibbed technical drawing pen
Rotring, *Hamburg, Germany – the 'Rapidograph'.*

Uranium, graphite-moderated, gas-cooled reactor
Marcoule Atomic Power Station, France – unsuccessful.

Xenon short-arc lamp
Dr Kurt Larché *and researchers at Osram GmbH, Germany – called the 'Wotan* XBO'.

1955
Air-to-air guided missile
USAF Hughes GAR-1 Falcon *for the*

Northrop Co.'s F-89 Scorpion fighterplane.

Anti-proton
Emilio Gino Segré, Owen Chamberlain, Clyde E Wiegand *and* Thomas Ypsilantis, *physicists at the University of California, Berkeley,* USA.

Automatic electric kettle (vapour-controlled)
Bill Russell *and* Peter Hobbs, *Dartford,* UK.

Bubble-oxygenator (cardiac)
Dr Clarence W Lillehei *and* Dr Richard A DeWall *at the University of Minnesota Hospital, Minneapolis,* USA.

Diego blood group system
M Layrisse, Arends *and* Dominguez, *Venezuela.*

Electric locomotive (remote radio-controlled)
SNCF, *France. It travelled from Paris to Le Mans at 80 mph.*

Felt-tip marker (refillable)
Esterbrook *of Birmingham,* UK – *called the 'Flo-Master'.*

Fibre optics
Dr Narinder S Kapany, *Imperial College, London, based on a discovery made 80 years before by* John Tyndall, UK *physicist.*

Graphic conversation terminal
Jay Wright Forrester *of Massachusetts Institute of Technology, Boston, with the 'Whirlwind' binary computer.*

Hovercraft
Christopher Sydney Cockerell, *electronics engineer with a small boatbuilding concern on the Norfolk Broads,* UK.

Hydropneumatic suspension (motor car)
Citroën, *France.*

Large radio telescope
Sir Bernard Lovell *at Jodrell Bank, Cheshire,* UK, *following five years of construction. Its 'dish' measured 250 ft in diameter.*

Lego® system of play
Godtfred Kirk Christiansen, *using interlocking plastic toy bricks as developed with his father,* Ole Kirk Christiansen, *carpenter and joiner of Billund, Denmark. 'Leg godt' (Danish) = play well. 'Lego' (Latin) = put together.*

Mendelevium
Albert Ghiorso, *University of California, Berkeley,* USA.

RNA (ribonucleic acid) synthesis
Severo Ochoa, *Spanish-*US *physician, Chairman of the Department of Biochemistry at New York University Medical College,* USA.

Spectacle lenses (unbreakable)
'Orma 1000' using CR-9 polymer.

Stereophonic tape-recordings
EMI *'Stereosonic' tapes.*

Submarine (nuclear-powered)
'Nautilus', under the direction of Captain Hyman G Rickover, US *Navy. On sea trials, 'Nautilus' cruised 60,000 miles on a single charge of fuel, using up no more than 8 lb of uranium.*

Super-aircraft carrier (100 jet aircraft)
'Forrestal', US *Navy.*

Tilting-rotor fixed-wing aircraft
Bell Aircraft Co.'s XV-3.

Tractor-mounted hydraulic digger-loader
J C Bamford Ltd, *Uttoxeter, Staffordshire,* UK.

Transoceanic telephone cable
American Telephone and Telegraph Company. *It was laid by the cableship 'Monarch' for a distance of 1950 nautical miles, from Clarenville, Newfoundland to Oban, Scotland.*

Ultrasound
Leskell, US, *to observe the heart.*

1956
Amniocentesis (routine procedure)
St Mary's Hospital, *Manchester,* UK.

Contact lens (plastic corneal lens)
Norman Bier, UK.

DNA biosynthesis
Arthur Kornberg, *Head of the Department of Microbiology at Washington University, St Louis,* USA.

'Dyform' (seven-wire and nineteen-wire pre-stressing strand)
the Steel Division of British Ropes Ltd, UK.

Flexible wing
Francis Melvin Rogallo, *aeronautical research scientist, and his wife Gertrude,* US.

Go-kart
Art Ingles, *Echo Park Road, Los Angeles, California.*

Human growth hormone (HGH)
Choh Hao Li, *Chinese biochemist working at the University of California, Berkeley,* USA. *Since used to wipe out dwarfism in children.*

Letraset
Fred McKenzie, *chemist, and* Dai Davies, *commercial artist,* UK, *who used the apparently simple principle of the child's waterslide transfer to develop a headline typesetting process. The Letraset business was founded in London in 1959. Two years later Letraset's Instant Lettering dry transfer process created a revolution in art-work and typographic layout.*

Merchant ship (gas-turbine/ electric)
'John Sergeant,' US *freighter.*

Motorcycle (desmodromic valves)
Fabio Taglioni, *chief designer, Ducati Meccanica* SPA, *Bologna, Italy.*

Plutonium power reactor (large-scale)
Calder Hall Atomic Power Station, Cumberland, UK, *one year after an experimental pressurised-water system station had gone into operation at Obninsk, near Moscow,* USSR.

Record-players (in automobiles)
Chrysler, USA.

Video-tape (commercial)
Charles Ginsberg, Charles Anderson, Ray Dolby, Fred Pfost *and* Alex Maxey, *engineers at Ampex Co., Redwood City, California,* USA. *The* VTR *spool began to be replaced by cartridges from 1969.*

Vitamin B12 (cyanocobalamin)
Karl August Folkers, *Merck and Co. Inc., Rahway, New Jersey,* USA.

1957

Air-to-surface missile
the US Bell GAM-63 Rascal, *as carried by Boeing* B-47 *Stratojet bombers,* USAF.

Air-turbine dental drill
spinning at 300,000 rpm.

Animal-carrying spacecraft
USSR. *A bitch, Laika, orbited the earth, supplied with food and oxygen, whilst her respiration and heart behaviour, telemetred, were transmitted back to earth. The satellite completed 2370 orbits before it re-entered the denser atmosphere and burned up.*

BCS theory (superconductivity)
John Bardeen, Leon N Cooper *and* J Robert Schrieffer, US *physicists.*

Cosmodrome
launchpad complex for Soviet manned and unmanned spacecraft, believed to be at Tyuratam in Kazakhstan, USSR.

Fibre-optic tubes (medical)
Dr Basil Isaac Hirschowitz, *South African physician at the University of Michigan,* USA, *for exploring the interior of the human body.*

Fortran (computer programming language)
John Backus *and a* research team working for IBM Corporation, *New York City, for use with the* IBM 704 *computer. A Fortran manual accompanied each computer sold.*

Glass ceramics
marketed as 'Pyroceram' in the USA *and in the* UK *as 'Pyrosil'.*

Ground effect (vehicles)
Louis Duthion, *France.*

Intercontinental ballistic missile (ICBM)
USSR – *called the* SS6 *(Sapwood), with*

a 6000-mile range.

Interferon (protein)
Dr Alick Isaacs *and* Jean Lindemann *(Swiss), biochemists at the National Institute of Medical Research, Mill Hill, London.*

Mechanised sorting of mails
British Post Office engineers, *Southampton,* UK. *The system of post-coding was initiated experimentally two years later, but did not come into complete nationwide use until 1974.*

Mössbauer effect (gamma radiation)
Rudolph Ludwig Mössbauer, *German physicist at the Max Planck Institute, Heidelberg University.*

Pacemaker (internal)
Dr Clarence Walton Lillehie, *surgeon at the University of Minnesota Hospital, Minneapolis,* USA, *with technical assistance from Earl Bakk, electronics engineer.*

Plumbicon TV camera tube
Philips Co., *Eindhoven, Netherlands.*

Polio vaccine (live virus)
Dr Albert Bruce Sabin, *microbiologist, New York, following Dr J F Enders' isolation of the virus in a test tube, some 9 years before, at the Children's Hospital, Boston. This vaccine was first adopted in the* USSR, *because medical authorities in the USA were more interested in other types of vaccine which proved, in the long run, far less effective.*

Pressurised water-cooled reactor
Westinghouse Corporation *of Pittsburgh, Pennsylvania,* USA, *for their Shippingport atomic power station.*

Satellite
Dr Sergei Pavlovich Korolyov *and a Soviet research team. Called 'Sputnik' (fellow traveller) I this 23 inch diameter metal sphere, travelling at a speed of 17,000 mph, orbited the earth 1400 times. It measured, among other things, the concentration of electrons in the ionosphere, before* returning to the lower atmosphere and burning up early in 1958.

Semi-synthetic penicillins
Messrs Doyle, Naylor, Rolinson *and* Batchelor *of the Beecham Research Laboratories, Brockham Park,* UK. *The 10-year research and development programme may have cost Beechams £12 million, but the sales of 'Broxil', 'Celbenin', 'Penbritin', etc. were to repay that investment many times over.*

Silicon-grown junction transistor
Texas Instruments, USA.

Steel-rolling mills (automatic thickness control)
Davy-Ashmore Ltd *with* United Instruments Ltd, *Sheffield,* UK.

305 RAMAC (random access method of accounting and control)
Ray Clifford Johnson *and an* IBM *research team,* USA, *who developed a magnetic disc storage system.*

Tilt-wing aircraft
Vertol Model 76, the VZ-2A, USA.

1958
'Airstrip' (microporous plastic medical dressing)
Smith & Nephew Ltd, *Hull,* UK.

ALGOL (computer language)
Swiss computer scientists.

All-transistor FM radio
Sony Corporation, *Tokyo, Japan.*

Banking credit card
Bank of America.

Communications satellite
SCORE *(signal communication by orbital relay), built and launched from the* USA.

Float glass process
Alistair Pilkington, *St Helens, Lancashire,* UK. *From 1952 until 1958 Pilkington developed this process, ignorant that it had been investigated but never tested by* William Heal *in 1902 and* H K Hitchcock *in 1905, both in the* USA.

1958

Geodesic dome (large scale)
Richard Buckminster Fuller –
*inventor-architect. The 384 ft
diameter Union Tank car dome at
Baton Rouge, Louisiana,* USA.
High-speed production of bread
Milling and Baking Industry
Research Centre, *Chorleywood,
Hertfordshire,* UK.
'Hula-hoop'
Richard P Knerr *and* Arthur K
'Spud' Melvin, *owners of Wham-O-
Manufacturing Co., San Gabriel,
California,* USA. *It was a tubular
plastic hoop with a choice of colours.*
**Intercontinental ballistic
missiles (greater range)**
USSR*'s* T-3 *followed in 1959 by the*
USA*'s Atlas* D.
**'Lexan', 'Makrolan', 'Merlon'
(polycarbonates)**
Farfabriken Bayer, *Germany, and*
General Electrics, USA.
LISP (computer language)
John McCarthy, *computer scientist,
Massachusetts Institute of
Technology, Cambridge,* USA *for
'artificial intelligence' research.*
Nobelium
Albert Ghiorso, *Lawrence Radiation
Laboratory, University of California,
Berkeley,* USA. *It is called 'Joliotium'
in the* USSR.
**Optical-wavelength maser
(laser)**
Charles A Townes, *physicist,
Columbia University, New York, with
his brother-in-law,* Arthur Leonard
Schawlow, *Bell Telephone
Laboratories.*
**Rotary blade electric lawn-
mower**
Wolf, *France.*
Stereophonic disc recordings
Audio Fidelity, US, *and* Pye, UK.
**Transistorised echo sounder
(marine)**
Major Richard Gatehouse, *formerly
of the Royal Radar Establishment,* UK.
*It was tested on his Lymington one-
design sloop 'Wavecrest II'. Two*

*models of the 'Hecta Mark I' appeared
at the 1959 London Boat Show.*
Unitary fibreglass automobile
Colin Chapman's *Lotus Elite,* UK.
Van Allen radiation belts
James Alfred Van Allen, *Professor of
Physics, University of Iowa,* USA.
*Based on geiger counter readings made
from the 'Explorer I' satellite.*
Wooden sailboard (prototype)
Peter Chilvers, *twelve-year-old
schoolboy at Hayling Island,* UK.

1959
**AG proton synchroton (high
energy particle accelerator)**
CERN *Laboratory for nuclear research
in Geneva, Switzerland, designed and
built with support from eleven
European countries.*
**Amphibious hydrofoil (gas-
turbine-engined)**
Lycoming Co., *under contract to the*
US *Army Ordnance Corps.*
**Animal radio-tracking
(prototype system)**
Messrs Le Munyan, White, Nybert
and Christian, US, *to track the wild
chipmunk.*
Anti-missile missile
*Nike-Zeus, test-fired from White
Sands, New Mexico,* USA.
**Artificial intelligence (AI)
computer program (to play
chess)**
Claude E Shannon, *Bell Telephone
Laboratories.*
**BWR (heavy-water-moderated
boiling water reactor)**
Halden *Atomic Power Station,
Norway.*
COBOL (computer language)
US *Conference on Data Systems
Languages.*
**Computer (performing eight-
data-processing tasks
simultaneously)**
Honeywell, US – *the 'Model 800'.*
**Computer-controlled chemical
plant**
Imperial Chemical Industries, UK.

Computer-controlled refinery
Texaco Oil Co., US – *for their Port Arthur plant.*

Full-sized hovercraft
Richard Stanton-Jones *and* research team at Saunders-Roe Ltd, *Cowes, Isle of Wight*, UK. *It was called the* SR.NI.

Fully simultaneous, large-scale computer system
Bull, *France – the 'Gamma 60'.*

Ice-breaker ship (nuclear-powered)
USSR. *The 440 ft 'Lenin' was capable of operation throughout the winter without refuelling.*

Identikit
Hugh C McDonald, *detective in the Los Angeles Identification Bureau. It was used by Sheriff Peter Pitchess of Los Angeles County Police to identify a criminal from McDonald's 500 master foils.*

Integrated circuit (phase-shift oscillator)
Jack S Kilby, *Texas Instruments, Dallas*, USA.

Ion engine (cesium contact)
Dr Alvin Theodore Forrester, *physicist, Rocketdyne*, USA.

Librium (drug)
Hoffmann LaRoche Pharmaceutical Co., *Switzerland.*

Lunar observation orbiter
USSR. *Called 'Lunik III', it circled the moon and photographed the far side, radioing pictures back to earth.*

Microwave point-to-point radio system
Pacific Great Eastern Railway, between Vancouver and Dawson Creek-Fort St John, British Columbia, Canada.

Morris Minor 'Mini'
Arnold Constantine Issigonis, *British Motor Corporation*, UK.

Motor car safety belt (production model)
Volvo, *Sweden.*

Plastic fishing nets
Smith & Nephew Plastics Ltd,

Harlow and Hull, UK.

PDP-1 (programmed data processor)
Ken *and* Stan Olsen *and* Harland Anderson *of the Digital Corp., Maynard, Massachusetts*, USA – *a small computer.*

Snow motor scooter
J Armand Bombardier, *Canadian mechanic – called the 'Ski-Doo'.*

Solid-state data-processing system
Olivetti, *Italy, for their 'Elea 9003'.*

Submarine (Polaris)
US *Navy's 'George Washington', capable of firing a Polaris missile, as developed two years previously, from beneath the water surface. It was built at and launched from Groton, Connecticut.*

Toughened float-glass car windscreens.

Transistorised computer
The IBM 1401.

Tunnel diode (commercial)
Sony, *based on work by* Leo Esaki, *Tokyo University, Japan.*

1960

Aircraft carrier (nuclear-powered)
USS *'Enterprise', 1101 ft long and capable of carrying 100 aircraft.*

Argon ion laser
D R Herriott, A Javan *and* W R Bennett *at Bell Telephone Laboratories, New Jersey*, USA.

Artificial ski slope
Len Godfrey *of Dendix Ltd*, UK, *manufacturers of industrial brushes in Vybak pvc monofilaments. The first 'Snowslope' was tested at Philbeach Hall, Earls Court, London.*

Balloon communications satellite
Echo 1, US.

Bathyscaphe (deep-sea)
Auguste Piccard, *Swiss physicist. His son* Jacques, *together with* US *Naval officer*, Lieutenant Don Walsh, *took the 'Trieste' to a depth of 35,800 ft*

into the Marianas Trench off the Pacific Island of Guam. *They observed the deepest ocean floor for half an hour through special plexiglass portholes.*

Bird radio-tracking system
Einar Eliassen, *Bergen, Norway, for mallard in normal flight.*

Closed-circuit TV camera as viewfinder for motion picture filming
David Samuelson, *cinematographer of Cricklewood, London.*

Computer (6000 additions or subtractions per second)
US Honeywell *'Model 400'.*

Computer-controlled production line
North Carolina Works of Western Electric, *Hickory, North Carolina,* USA.

Digital flight simulator
University of Pennsylvania, USA, *and manufactured by the Sylvania Corp. as* 'UDOFT'.

Electron bombardment engine
Harold R Kauffmann, *space propulsion engineer,* NASA, *Cleveland, Ohio,* USA.

Fibre-tip pen
Japan Stationery Co., *Tokyo – called the 'Pentel'.*

Intercontinental ballistic missile (8000-mile range)
US *Minuteman I.*

Ion source (implementation gun)
J Harry Freeman, *nuclear chemist,* UK *Atomic Energy Research Establishment, Harwell, Oxfordshire. Initially for isotope separation, it was developed to make materials tougher and more corrosion-resistant.*

Missile launch warning satellite
USAF.

Moulton bicycle
Alex Moulton *at the small research establishment of his family estate in Wiltshire,* UK. *By 1965 major cycle makers in seven countries had taken licences for the manufacture of the Moulton bicycle with its F-shaped*

frame and small wheels.

Nitinol (memory-metal)
William Buehler, *engineer at Silver Spring Research Laboratories,* USA.

Pulsed ruby laser
Theodore H Maiman, *physicist, at Hughes Research, Malibu, California,* USA.

Re-locatable building module (loda-strut-leg system)
Donald Shepherd, *building contractor, York,* UK. *To produce working facilities on building sites. Called the 'Portakabin'. 50–60,000 currently in versatile use worldwide.*

Turbojet-engined car
Dr Nathan Ostich, US. *His 'Flying Caduceus', when tested on the Bonneville Salt Flats, Utah, was found mechanically unsuccessful.*

Typesetting by computer
Imprimérie Nationale, *Paris.*

VTOL (vertical take-off and lift) aircraft
Frank Taylor *and a* design team *at Short Brothers & Harland at Sydenham, Northern Ireland. The* SC-1*'s test pilot was Tom Brooke-Smith and the powerplant was the Rolls-Royce* RB108 *engine.* VTOL *tests had been carried out some six years before with the Rolls-Royce-engined 'Flying Bedstead'.*

Weather satellite
NASA's TIROS-1.

1961
Back-pack aero-engine
Wendell F Moore, *engineer at the* US *Bell Aircraft Corporation. The* SRLD *(small rocket lift device), strapped on the back of Harold Graham, made its first untethered flight.*

Cargo and passenger ship (nuclear-powered)
'The Savannah', US, *21,800 tons with a passenger capacity of 60. It proved too costly and was laid up in 1967.*

Computer time-sharing experiment
Ferdinand José Corbató,

Computation Centre, Massachusetts
Institute of Technology, Cambridge,
USA, for exploitation with the IBM
'709' model.
**Computerised algorithmic
music**
Pierre Barbaud and colleagues at
Compagnie de Machines Bull, Paris.
Electric toothbrush
Squibb & Co., New York.
**Electronic calculator
(commercial model)**
Sunlock Comptometer Ltd, UK –
called the 'Anita'.
**Flexible skirt (hovercraft
device)**
C H Latimer-Needham, UK.
'Flying saucer'
Avro-Avrocar VZ-9V, Canada.
Powered by three Continental J69
turbojets, it was abandoned as
uncontrollable.
**Gas-assisted gas-laser cutting
tool**
British Welding Institute, marketed
by the British Oxygen Co.
Golf ball electric typewriter
'IBM 72', US.
**Hatchback estate car (upward-
opening rear door)**
Renault 4, France.
**'Kenyapithecus Wickeri' (upper
jawbone)**
Dr Louis Seymour Bazett Leakey,
prehistoric archaeologist. The
jawbone, found at Fort Ternan in
Western Kenya, was dated to 14
million years old.
Lawrencium
Albert Ghiorso, nuclear scientist,
Lawrence Radiation Laboratory,
University of California, Berkeley,
USA.
**Magnetic ink character
recognition**
Honeywell, US, to read banking
computer print-outs.
Man-carrying spacecraft
USSR. Travelling in the 4.5 ton 'Vostok
I', Flight Major Yurij Alexeyevich
Gagarin, aged 27, completed a single

'Vostok I', cosmonaut-carrying spacecraft

orbit of the earth, making a safe
landing in the USSR one hour and 48
minutes later.
Nuclear bomb (50 megatons +)
exploded by the USSR.
Observation/'spy' satellite
USAF.
Self-twist yarn spinner
David Henshaw, physicist, Textile
Division of the Commonwealth
Scientific and Industrial Research
Organisation, Geelong, Victoria,
Australia.
Space center
Johnson Space Center, Houston,
Texas, USA. Flight controllers at
Mission Control were subsequently
equipped with as many as 171 consoles
including 133 TV cameras and 557
receivers.
Stereophonic radio broadcasts
Zenith and General Electric, USA.
'Stretch' (computer system)
IBM. 150,000 transistors enabled it to
perform 100 billion computations per
day, using an Octet (eight-bit) system.
**Transistorised video-tape
recorder**
Sony SV-201, Tokyo, Japan.
'US Patent No. 3,000,000'
Kenneth R Eldredge for the General
Electric Co., New York: automatic
reading system.
Valium (drug)
Hoffmann-LaRoche Laboratories,
Switzerland.
**Zone-toughened car
windscreens**
Triplex-Glass, UK.

1961

1962
Aérotrain (tracked hovercraft)
J Bertin *and* P Guienne *of Bertin et Cie, France. In 1974, their vehicle achieved a speed record average of 425 kph.*
AG electron synchroton
Massachusetts Institute of Technology, *Cambridge, with* Harvard University, *supported by the* US *Atomic Energy Commission – known as the Cambridge Electron Accelerator.*
Air-bearing, air-turbine dental drill
to run at 800,000 rpm.
'Allopurinol' (drug)
Dr George H Hitchings *and the Burroughs Wellcome research staff, to give relief from gout. It was first administered to gout patients at Duke University Medical Center at Durham, USA. 'Allopurinol' was Hitching's fourth great medical drug discovery – 'Daraprim', '6-MP' and 'Azathioprine' being the previous three.*
Amphicar
Hans Trippel, *Germany. It was developed from his 'Alligator' prototype of two years earlier. 3000 Amphicars were manufactured and sold by 1965.*
Autoflare (automatic landing equipment)
developed with BOAC, UK, *for the Vickers* VC10 *airliner.*
Ballistic-missile submarine
US Navy's *'Ethan Allen'.*
Computer simulation languages
US Rand Corp.'s SIMSCRIPT and IBM's GPSS.
Computerised 'real time' flight reservation system
IBM *in conjunction with American Airlines – called 'Sabre'.*
Experimental atomic power station (fast breeder reactor type)
UK *Atomic Energy Authority, Dounreay, Caithness, Scotland.*

Gallium-arsenide semiconductor laser
R N Hall, *General Electric Co., Schenectady, New York,* USA.
Gamma-ray sterilisation (medical and surgical products)
UK *Atomic Energy Authority, as suggested by* G S Murray.
Laser microbeam
Marcel Bessis *and* a team of Parisian biologists, *producing a 2.5 micron diameter spot for work in cell biology.*
LED (light-emitting diode)
Nick Holonyak Jr, *Advanced Semiconductor Laboratory, General Electric, Syracuse, New York,* USA.
Lunar-reflected optical maser beam
Bell Telephone Laboratories, USA.
Magnox (magnesium non-oxidising) atomic power stations
Berkeley, Gloucestershire and Bradwell, Essex, both UK.
Micro-television (5 inch screen)
Sony, *Tokyo, Japan – called the* TV5-303.
Minicomputer
Digital Corporation *of Maynard, Massachusetts,* USA – *called the* PDP-5.
Muon neutrino (elementary particle)
Leon Max Lederman *et al. at Columbia University, New York,* USA.
Passenger hovercraft
Saunders-Roe Ltd, *Cowes, Isle of Wight,* UK. *The* SR.N2 *had a carrying capacity of 70 passengers.*
Primitive industrial robot
Unimation Inc., *Dambury, Massachusetts,* USA.
Silicon integrated circuits
Texas Instruments, USA – *complex circuit functions in a single chip.*
Transatlantic satellite transmission
American Telephone and Telegraph Co., *using 'Telstar' to transmit an image from Maine,* USA *to Cornwall,* UK.

X-ray astronomy
using rocket-borne equipment to identify the 'Crab Nebula'.

1963
Anti-Xi-Zero (atomic particle).
Cassette audio-tapes
Philips, *Eindhoven, Netherlands.*
Computerised stylometric analysis
Rev. Andrew Q Morton *at the University of Edinburgh,* UK, *to discover and authenticate the authorship of the Pauline Epistles.*
Etorphine (M-99)
Dr Kenneth W Bentley *and* D G Hardy *of Reckitt & Sons Ltd, Hull,* UK.
Geosynchronous satellite
NASA – *called* 'SYNCOM 2' *and* 'SYNCOM 3'.
'Liberator' concept
automatic translation of computer programs from IBM*'s Model* 1401 *to Honeywell's Model* 200.
Motor car (Wankel engine)
NSU *'Wankel Spyder', Germany, although large-scale production came in 1968 with the 'Mazda 110S', Japan.*
Navigation satellite
NASA*'s 'Transit'.*
Quarks
Murray Gell-Mann, *theoretical physicist, California Institute of Technology, Pasadena,* USA.
Quasars (quasi-stellar radio sources)
Marten Schmidt, *Dutch astronomer, Hale Observatory, Palomar, California,* USA.
Rotary lawn-mower with hover
Flymo, UK.

1964
'A–Z System' (new knitting yarns from man-made fibres)
Linen Industry Research Association, *Lisburn, County Antrim, Northern Ireland, with* McCleery and L'Amie Ltd, *Belfast.*

Acrylic paints
Reeves Ltd, UK – *called 'Acrylic Vinyl Co-Polymer Paints'.*
Airflow ventilation (via automobile dashboard)
Ford Motor Co, USA.
BASIC (beginner's all-purpose symbolic computer instruction code)
Thomas E Kurtz, *teacher, and* John G Kemeny, *mathematician, Dartmouth College, New Hampshire,* USA.
Carbon fibre
RAF Farnborough, UK, *following 28 years' research and development. It was manufactured in the* UK *by Courtaulds Ltd as 'Grafil' and in the* USA *by Hercules Inc.,* ISD, *Utah.*
Computer-aided graphic design system
IBM 2250
Crew-carrying spacecraft
USSR. *The three-cosmonaut crew remained in 'Voshkod I' for sixteen orbits (or 24 hours) to test collaboration on board.*
Fibreglass sailboard
S Newman Darby *and his wife* Naomi, *at Falls, Pennsylvania,* USA. *The plans were published in 'Popular Science' magazine soon after.*
Fibrillated polypropylene
British Ropes Ltd *with* Plasticisers Ltd *for rope and cordage manufacture.*
Home-use, all-transistor, video-tape-recorder
Sony, *Tokyo, Japan – called the* 'CV-2000'.
Hovercraft (production model)
British Hovercraft Corporation, *Cowes, Isle of Wight,* UK. *The* SR.N5 *45 were built and in service by 1968.*
'IBM System 360'
Gene M Amdahl, M Blaun *and an* IBM research team *at New York – the high-volume, automatic, microminiature production of Solid Logic Technology (*SLT*), semiconductor, ceramic module circuits.*

Laser eye surgery
H Vernon Ingram, *ophthalmic surgeon*, USA, *using a* 'NELAS'.
Laser gun (prototype)
USA.
Mellotron
the Bradley brothers (Norman, Frank *and* Leslie) *of Birmingham*, UK. *The* BBC's *Mellotron Mark II of the following year could produce 1260 special effects.*
Rhesus haemolytic disease (treatment for)
simultaneously by Doctors Clarke, Sheppard *and* Finn *at Liverpool University Medical School*, UK, *and* Doctors Gorman, Freda *and* Pollock, *sponsored among others by the Ortho Pharmaceutical Foundation, New York. The first trial of the anti-Rh gamma globulin vaccine was on Dr Gorman's own sister-in-law. By 1967 it was being administered to women en masse worldwide.*
3-D laser-holography
Emmett Norman Leith *and* Juris Upatnieks *at the Willow Run Laboratory, University of Michigan,* USA.
Satellite navigation
US Government *for use by their* 'Polaris' *submarines*.
Unnilquadium
Albert Ghiorso *et al., Heavy Ion Linear Accelerator, Lawrence Radiation Laboratory, University of California,* USA. *The American name is Rutherfordium; the Soviet name is Kurchatovium.*
Voltage-control music synthesiser
Dr Robert Moog, US *engineer*.

1965
Airliner (automatic landing)
Paris-London BEA *de Havilland* 'Trident' *at Heathrow Airport, London.*
Artificial tooth
researchers at the University of Michigan School of Dentistry, USA, *to record and broadcast dental pressures. It contained six radio transmitters, 28 electronic components and two rechargeable batteries.*
Carbon fibres as prosthetic material
Henry G. de Laszlo, *physicist, Englefield Green,* UK.
Cartridge audio tapes
Learjet, *Netherlands*.
Computer-carrying spacecraft
NASA's 'Gemini 3', *as piloted by astronaut Commander Virgil Grissom,* US.
Cosmic microwave background radiation
Arno Allen Penzias, *astrophysicist, and* Robert Woodrow Wilson, *radio astronomer, Bell Laboratories, Holmdel, New Jersey,* USA.
Global communications satellite
NASA's 'Early Bird', *the first of the five* 'Intelsats'.
Hydraulic mobile and truck cranes (synchronised telescopic jibs)
Coles Cranes Ltd, *Sunderland,* UK.
'Kevlar'
Stephanie Kwolek *and* researchers *at Du Pont de Nemours,* USA. *By 1969, as* PRD-49, *this aramid was available to select* US *government departments. By 1973 it had been marketed under the computer-chosen names of* 'Kevlar 49' *and* 'Kevlar 29'.
Medium density fibreboard
USA – *dry-formed panel manufactured from lignocellulosic fibres bonded with a synthetic resin.*
Minicomputer (mass-produced)
Digital Corporation *of Maynard, Massachusetts,* USA – *called the* 'PDP-8'.
Mini-skirt (woman's dress)
Mary Quant, *fashion designer, King's Road, Chelsea, London.*
'Multics' (computer time-sharing system)
General Electric *for project* 'Mac' *at both the Massachusetts Institute of Technology and Bell Telephone*

Laboratories, USA.

Natural gas (North Sea)
British Petroleum, UK. *Natural gas was piped ashore from the West Sole Field for the* UK's *domestic gas system by 1967.*

Radial tyre (asymmetric tread)
Michelin et Cie, Clermont-Ferrand, France. *Called the 'XAS'.*

Rubella vaccine
Dr Paul D Parkman *and* Dr Harry M Meyer Jr *at the Division of Biologic Standards of the National Institute of Health in Bethesda, Maryland,* USA.

Scanning electron microscope
Engineering Department of the University of Cambridge *with* Cambridge Instruments Ltd, UK. *Called the 'Microscan* SEM', *it gave a depth of field 300 times greater than any previous optical or electron microscope.*

'UK Patent No. 1,000,000'
Ernest Benson, Garforth, near Leeds, UK: *'improvements relating to extensible loft ladders'.*

Word processor
IBM – *magnetic tape storage system linked to an electronic typewriter.*

1966
Integrated circuit radio
Sony, Tokyo, Japan – *called the* 'ICR-100'.

International chess match by computer
Professor John McCarthy's *computer, 'Stanford University', versus Abram Alikhanov's computer, Russian 'Institute of Theoretical and Experimental Physics'. The* USSR *won 3–1.*

Josephson effect device
Brian David Josephson, *physicist, University of Illinois,* USA.

Laser automobile design system
Hitachi Ltd *and* Mitsubishi Electric Corp., Japan.

Linotron
Mergenthaler-Linotype Co., US, *in collaboration with* CBS *Laboratories to develop an electronic machine using a cathode ray tube to form letters at rates over 1000 characters per second. The initial prototype for a cathode-ray-tube machine had been developed 20 years before by* E Dinga.

Noise reduction system (audio recording and transmission)
Dr Ray M Dolby, US-*born electronics engineer and physicist in South-west London,* UK. *Called 'Dolby A', it was first used professionally by the Decca Record Co.*

Racing car with aerofoils
the Chaparral 2E, US, *as driven by Jim Hall at the Bridgehampton Can-Am event.*

Skateboard
California, USA.

Supertanker
the 205,000 ton 'Idemitsu Maru', built in Japan.

Telephonic optical-fibre link (experimental)
Dr Charles Kao *and* Dr George Hockham *at Standard Telecommunications Laboratory, Harlow, Essex,* UK.

1967
Bubble memory (prototype)
A H Bobeck *and researchers at Bell Telephone Laboratories, Holmdel, New Jersey,* USA.

'Cache'/Buffer memory circuit card
IBM *System/360 for their Model 85 computer.*

Computerised electronic fuel-injection system (motor car)
Robert Bosch GmbH, *Stuttgart, Germany.*

Dynamic memory cell
IBM.

International computer time-sharing
General Electric, US, *between Toronto, Canada and London,* UK. *It was later extended to Sydney, Paris, Milan, the Hague, Brussels, Cologne, Copenhagen and Stockholm.*

1967

Laser-lancet (medical)
D R Herriott, E I Gordon, H A S Hale *and* W Gromnos *of Bell Telephone Laboratories, New Jersey,* USA. *Called the 'Light Knife', it was used to cut or cauterise a wound with minute precision.*

Levodopa (drug)
George C Cotzias, *Greek physician, to control the horrifying symptoms of Parkinson's disease.*

Louma crane (cinematography)
Jean Marie Lavelou *and* Alain Masseron, *Paris. It was improved and marketed by Samuelson's of Cricklewood, London.*

Magnetic-motor train
Pueblo, Colorado, USA.

Precision-surveying laser
used during the construction of tunnels for the Bay Area Rapid Transit Railway, San Francisco, USA.

'Raudive voices' (communications from the dead)
Konstantin Raudive, *Latvian psychologist, living in Germany. He used a tape recorder to register 'white noise' from either a radio or crystal diode set with a very short aerial. Playback appeared to reveal 'voices'.*

Rocket car (liquid-fuelled)
Reaction dynamics 'X-1' dragster, designed and built at Milwaukee, Wisconsin, USA.

Satellite docking
USSR, *with automatic docking of the unmanned 'Cosmos 186' and 'Cosmos 188' satellites.*

Spinal meningitis vaccine
Doctors Artenstein, Gotschlich *and* Goldschneider, *working for the* US *Army at the Walter Reed Institute.*

Tidal-electric power project
the Rance Plant at the Gulf of St Malo, Brittany, France.

Tobiscope
Mikhail Gaikin, *Soviet surgeon in Leningrad, together with* Semyon Kirlian, *for the precise location of acupuncture points.*

1968

Carbon dioxide gas laser.

Electronic signature verification
a team of psychologists and electronic engineers *working for three years at* the National Physical Laboratories, *Teddington, Middlesex,* UK. *The prototype machine, to read human handwriting, was called 'Cyclops 3'. Commercial versions included 'Securisign' from the Transaction Security division of Analytical Instruments Ltd,* UK.

Grass-skis
Kurt Kaiser, *knitting-machine manufacturer, Stuttgart, Germany, who pioneered the sport of running down grassy hillsides on 1 ft long 'caterpillar-track' skis.*

Haemoglobin molecule structure (complete)
Dr Max Ferdinand Perutz, *Director, Medical Research Council's Molecular Biology Laboratory, Cambridge,* UK, *following 30 years' research – with assistance from his colleague* Dr John Kendrew.

Jumbo helicopter
Mikhail Mil, USSR: *the four turboshaft-engined Mil-V12 'Homer'.*

Lithium niobate (holographic storage material)
Bell Telephone Laboratories, *Holmdel, New Jersey,* USA.

Long-life, high-output argon and krypton laser
Shibaura Electric Co. Ltd, *with the Japan Central Research Laboratory, Tokyo.*

Pulsars
Dr Jocelyn Bell Burnell *at the Mullard Radio Astronomy Observatory, Cambridgeshire,* UK.

SGHWR (steam-generating heavy water reactor)
100 MW *Power Station at Winfrith, Dorset,* UK.

Submersible (glass-fibre construction)
Vickers Oceanics Ltd, *Barrow,* UK.

Supersonic airliner
*the Soviet-built Tupolev Tu-144, with
a maximum speed of Mach 2.37 and a
passenger capacity of 130. The Anglo-
French 'Concorde', designed and built
by British Aerospace with
Aerospatiale, first flew in 1969 and
soon reached Mach 2.0. 'Concorde' has
proved more viable commercially.*
Tractor (hydrostatic steering)
David Brown Tractors Ltd, UK.
**Trinitron (colour television
picture tube)**
Sony, *Tokyo, Japan – called the* 'KV-
1310'.

1969
**ALSEP (Apollo lunar scientific
experiment package)**
NASA*'s 'Apollo 12' astronauts, Conrad
and Bean, deployed six scientific
experiments each with a nuclear-
powered battery giving operational
equipment life of one year on the lunar
surface.*
**Antibodies (chemical molecular
structure of)**
Rodney Robert Porter, *Professor of
Biochemistry at Oxford University,*
UK *and* Gerald Maurice Edelman,
*Professor of Biochemistry at
Rockefeller University, New York,*
USA.
**Command and service lunar
modules**
NASA *with aerospace companies led by*
North American Rockwell *and*
Grumman Aircraft. *They were used
by the 'Apollo 11' astronaut team for
their landing on the moon, and for five
subsequent landings.*
Digital avionics flight computer
Honeywell Inc., *for the McDonnell-
Douglas* DC-10 *airliner.*
Gamma-ray astronomy.
**Gear wheels (carbon-fibre-
reinforced plastic)**
British Rail Engineering Ltd, UK.
Jumbo-jet airliner
Joe Sutherland *and a* design team *at*
Boeing, Seattle, Washington. Called

'Apollo' astronauts' spacesuit

*the Boeing 747, it made its first test
flight from Paine Field, watched by
PanAm Airways officials who had
already ordered 25 such 'Jumbo's.
Passenger capacity was 385-500.*
**Laser-ranging retro-reflector
(part of ALSEP)**
*Able to measure the distance of the
earth to the moon to an accuracy of 1.5
miles.*

Manned spaceflight docking machinery
Soviet spacecraft 'Soyuz 4' and 'Soyuz 5', followed by the 37-minute spacewalk transfer of two space-suited cosmonauts from one craft to the other.

PASCAL (computer language)
Professor Niklaus Wirth, *computer scientist, ETH, Zurich, Switzerland.*

Twin-engined offshore catamaran
James Beard, Clive Curtis *and* Chris Hodges *of Twickenham, Middlesex,* UK – *called 'Volare II'.*

VTOL fighter-plane (vectored lift)
Hawker Siddeley Aviation Ltd.
Called the 'Harrier', it was developed during the 1960s with the Hawker 'P1127' and the 'Kestrel'.

Video-cassette system (plastic-boxed)
Sony, *Tokyo, Japan – called the 'U-Matic'.*

1970
Airfoilcraft
VFW-Fokker GmbH, *Germany.*

'AI' computer programs (for understanding language)
Roger C Shank *of Yale University,* USA – *called* 'MTRANS', 'ATRANS' *and* 'ATTEND'.

Bar code (computer-scanned binary signal code)
simultaneously by Monarch Marking *of Dayton, Ohio,* USA *with 'Codebar' for retail trading, and* Plessey Telecommunications Ltd *of Poole, Dorset,* UK *for industrial use.*

Floppy disc (magnetic storage device for computer systems)
IBM.

Laser/hologram art exhibition
Finch College Museum of Art, New York. It was called 'N-Dimensional Space'.

Lunar vehicle (remote-controlled)
USSR's *'Lunokhod I', which travelled 6½ miles on gradients of up to 30 degrees*

in the Mare Imbrium and remained functional for almost one year.

Noise reduction system (cassette tapes)
Dolby B-type Encoder 320, UK.

Oil (North Sea 'Forties' field)
discovered by British Petroleum's drilling barge, 'Sea Quest'.

Overhead travelling crane (linear-induction-powered)
Linear Motors Ltd, *Loughborough, Leicestershire,* UK, *based on 15 years' research and development by* Professor Eric Roberts Laithwaite, *Heavy Electrical Engineering Department, Imperial College of Science and Technology, London.*

Relational data base
Dr Edgar Codd *and an* IBM research team.

SHRDLU (AI computer program)
Terry A Winograd *of the Massachusetts Institute of Technology, Cambridge,* USA, *to simulate actions/responses of a person in a 'Blocks World'.*

Sialon
Professor K H Jack *at the University of Newcastle-upon-Tyne and* Y Oyama *of Tokyo, Japan. It was developed in subsequent years as an engineering ceramic by* Lucas Research Centre, *Solihull,* UK.

Sixteen-bit minicomputer
Digital Equipment Corp. – *called the* 'PDP-11'.

Synthetic lubricant for supersonic aircraft engines
British Petroleum Ltd *with* Geigy (UK) Ltd.

SYSTRAN (computerised translation system)
Dr P Thomas. *It was adopted by the European Economic Community in 1981.*

Unnilpentium
Albert Ghiorso *et al., Lawrence Radiation Laboratory, University of California,* US. *The American name is Hahnium; the Soviet name is*

Nilsbohrium.
Videodisc (black and white)
Decca, UK, *and* AEG Telefunken,
Germany – called 'Teldec'.
**Windsurfer (commercial
sailboard)**
James Drake *of Santa Monica,
California, and* Hoyle Schweitzer *of
Pacific Palisades, California.*

1971
**'Anorthosite' (crystal moon
rock)**
Pilot-Astronaut David R Scott,
commander of NASA's *'Apollo 15,'
during a 28 km drive in the Lunar
Rover Vehicle, when he collected 28 kg
of rock samples.*
**Earth-orbiting space station
(USSR)**
*'Salyut', visited by cosmonauts from
the 'Soyuz 11' spacecraft.*
LCD (liquid-crystal display)
Hoffmann-LaRoche, *Switzerland.*
Micro-processor
Doctors Robert Norton Noyce *and*
Gordon Moore, US. *Launched by the
Intel Corporation, Santa Clara,
California, USA, it was called the 'Intel
4004'.*
Orbital engine
Ralph Sarich, *engineer, Australia*
PROLOG (computer language)
Alain Comerauer *and* Philippe
Roussel, *French researchers at
Strasbourg University.*
Quadraphonic disc recordings.
Quartz digital watch (LED)
George Theiss *and* Willy Crabtree,
US *engineers, Dallas, Texas – called
the 'Pulsar'.*
**Shirt-pocket radio pager (tone
and voice)**
Motorola, US.
Sony memory transistors
Sony, *Tokyo, Japan.*

1972
Advanced Passenger Train
British Rail's Research Centre *at
Derby, UK. Its top speed was 152 mph.*

**Anti-matter particles (anti-
helium 3)**
Professor Yuri Dmitriyevich
Prokoshkin *and a research team at
the atom-smashing particle accelerator
at Serpukhov, Armenia, USSR.*
Artificial hip
John Charnley, *British orthopaedic
surgeon at the Wrightington Hospital,
Wigan, Lancashire, UK, using high-
density polyethylene as the
constructional material for the ball and
socket.*
**Auto-focusing still camera
(commercial)**
Polaroid SX 70, *using an ultra-sonic
signal.*
Black holes (proven)
*Professor Robert L F Boyd and his
team at University College, London,
using X-ray telescopes on board the 2
ton* US *orbiting astronomical
observatory No. 4. They were
discovered in the binary-star X-ray
source, Cygnus X-1.*
**Earth Resources Technology
Satellite**
NASA's ERTS I, *later renamed* LANDSAT
I *and followed by* LANDSAT 2 (1975)
and LANDSAT 3 (1978). *They were used
to make a world atlas, taking
photographs from a height of 560
miles.*
Electronic pocket calculator
simultaneously by Jack St Clair Kilby,
with J D Marriman *and* J H Van
Tassel *of Texas Instruments, Dallas,*
USA, *and* Clive Marles Sinclair *of
Sinclair Radionics Ltd, Cambridge,*
UK.
Enkephalin (brain chemical)
Dr John Hughes *and* researchers *at
the Drug Research Institute of
Aberdeen University, UK. It was
isolated, purified and analysed and
finally synthesised during the
subsequent two years.*
**Fly-by-wire (air navigation
system)**
McDonald-Douglas' F4 *'Phantom II'
was the first to use this system.*

1972

Minehunter ship (glass-fibre)
150 ft HMS 'Wilton,' *built by Vosper-Thornycroft, Portsmouth,* UK.

Outer solar system probe
NASA's *'Pioneer 10', which was still transmitting information about spatial conditions 13 years later, having left the solar system in 1983.*

Polypropylene motor car bumpers
Renault 5, France.

Programmable hand-held calculator
Hewlett-Packard, *Palo Alto, California – called the* 'HP 35'.

Satellite-tracking of wildlife
Frank *and* John Craighead, US, *using a 'Nimbus 3' weather satellite to track an elk for 28 days.*

'SLURP' (self-levelling unit for the removal of pollution)
Esso Oil Corporation, USA.

Telepen
George Sims, *physicist, of Friern Barnet, Middlesex,* UK. *A bar-code light pen reader, it was manufactured two years later, for use particularly in libraries and factories.*

Videodisc (laser-read)
Philips, *Eindhoven, Netherlands. It was marketed in 1980 (*USA*) and 1982 (Europe).*

Video game
Noland Bushnel, *28-year-old* US *student. Called 'Pong', it enabled him to form his own company – Atari.*

1973

BMX (bicycle-motor-cross)
Marvin Church Senior *of California* for his son, Marvin Jr. *The* BMX *frame was marketed the following year by Webco, motorcycle accessory manufacturers of Southern California.*

Computerised axial tomography X-ray scanner (CAT)
Allan Macleod Cormack, *South African physicist, Tufts University, Medford, Massachusetts,* USA, *and* Godfrey Newbold Hounsfield, *Central Research Laboratories,* EMI,

Hayes, Middlesex, UK. Professor I Isherwood *of Manchester University developed its medical implications during initial clinical trials at Atkinson Morley's Hospital, Wimbledon. Two years later* EMI *produced a fast whole-body* CAT *scanner.*

DNA (recombinant)
Paul Berg, Stanley N Cohen *and* Annie C Y Chang *of Stanford University, California, and* Herbert W Boyer *with* Robert B Helling *of the University of California, San Francisco,* USA.

Earth-orbiting space station (US)
Nearly 26,000 scientists *and other workers were deployed on preparations for* NASA's *'Skylab' mission. It was subsequently visited by three separate Apollo astronaut crews.*

Extra-corporeal shockwave lithotripter
Professors Eisenberger *and* Chaussy *in the Department of Doctor Egbert Schmidt, Munich, West Germany. Built by Dornier GmbH of Friedrichshafen. First used in Munich in 1980 for the disintegration of kidney stones.*

Fast-reacting photochromatic ophthalmic glass
Optical Division of Chance-Pilkington, *St Asaph, Clwyd,* UK.

Flying saucer (nuclear-powered)
British Rail – *patented but never built.*

Gamma-ray hunting satellite
NASA *Special Astronomical Satellite –* SAS-2.

Microcomputer
Trong Truong, *Director of* RE2 Co., *France – called the 'Micral'.*

Motor car headlamp wipers
Saab 99, as developed by Robert Bosch GmbH, *Stuttgart, Germany, in 1967.*

Sailing dinghy (injection-moulded polypropylene)
Ian Douglas Ben Proctor, *yacht*

designer, UK, *with* Rolinx Ltd *of Manchester,* UK *and* J C Dunhill Boats *of Basingstoke, Hampshire, using* ICI *propathene. Called the 'Topper', 21,000 dinghies were produced by this system in a very short time.*

Sears Tower (skyscraper)
the headquarters of Sears, Roebuck & Co., Chicago, Illinois, USA. *It is 1,454 ft high with 110 storeys. The building's population of 16,700 is served by 103 elevators, eighteen escalators and 16,000 windows.*

Skateboard (urethane plastic wheels)
Frank Nasworthy, *Californian skateboard enthusiast.*

Sydney Opera House
Jørn Ųtzon *et al., following 16 years' design and construction by many various companies.*

Thermographic duplicating machine
Rapid Data Ltd, UK.

Turbocharged production motor car
BMW, *Munich, Germany – called the 2002 Turbo.*

1974
ARECIBO message
Professor Frank Donald Drake, *astronomer,* US. *Using a radio telescope in Puerto Rico, he sent out a powerful series of radio pulses to communicate with extra-terrestrial intelligences elsewhere in the universe. So far, no replies . . .*

18,000-lines-per-minute, non-impact printing system
Honeywell Inc., USA.

Electronic memory credit card
Roland Moreno, *self-taught engineer, France.*

Laser-engraving of printing plates
using the Pagitron System of Optronics Inc., with the Laserplate of Lasergraphics Inc., for Gannett Newspapers Inc., USA.

'Mailgram'
NASA's *'Westar' I and II satellites for Western Union, the* US *postal and telegraph service.*

Motorboat (Kevlar construction)
Fabio Buzzi, *Italy.*

Motorcycle (Wankel-engined)
Suzuki, *Japan.*

NRDC 143 and 161 (pesticides)
research team at Rothamsted Experimental Research station *at Harpenden, Hertfordshire,* UK. *These highly active, safe insecticides are 1000 times more potent than* DDT.

'Nodding boom' (wave-power device)
Stephen Salter, *Edinburgh,* UK.

Passenger hydrofoil (waterjet-propelled)
Marine Systems Division of Boeing Aircraft Corporation, *Seattle, Washington,* US. *The Jetfoil had a carrying capacity of 250–400 passengers. The first Jetfoil, 'Madeira', went into operation between Hong Kong and Macao in 1975.*

Psi (atomic particle)
Professor Burton Richter, *physicist, at the Stanford Linear Accelerator complex, California, and* Professor Samuel Chao Chung Ting, US *particle physicist of Brookhaven National Laboratory, New York – using a colliding beam machine called* SPEAR.

'Sensurround' (cinema)
Waldo O Watson *and* Dick Stimf *of Universal Pictures,* US.

Sony Mavica system
Sony, *Tokyo, Japan. It uses magnetic video card for recording and playback.*

Teletext (computer-based television information service)
the engineering departments *of the* BBC *and the Independent Broadcasting Authority, using redundant space in* TV *waveform.*

Unnilhexium
Glenn Theodore Seaborg *and* Albert

Ghiorso *at the Lawrence Berkeley Laboratory, University of California,* USA.

1975

Carbon brakes (commercial aircraft)
Dunlop Ltd, British Aerospace *and* British Airways, UK.
Computerised astrology
Esther Blumenthal's *'Astrocan 2000'*.
Condeep
Aker *and* other Norwegian contractors *at Arendal, Stavangar, Norway – a concrete production platform for the* UK *sector of Mobil Oil's 'Beryl A' offshore oilfield. A similar version was built the following year by* McAlpine Seatank, UK.
Docking module assembly (Apollo-Soyuz test project)
both NASA *and* Soviet engineering teams *and* aerospace companies.
Electronic sewing machine
Singer, USA.
High-speed train
British Rail's Research Centre *at Derby,* UK. *Its top speed was 125 mph.*
Home videotape systems
Sony's *'Betamax' and* Matsushita/ JVC's *'Video Home System' (*VHS*), both of Japan.*
Hybrid cells (half animal/half plant)
Professor Jack Lucy, *Royal Free Hospital, London, and* Ted Cocking *of Nottingham University,* UK.
Laser cell-sorting machine
Biophysics and Instrumentation Group, Los Alamos Scientific Laboratory, *New Mexico – to detect cervical and vaginal cancer.*
Mass storage system (magnetic tape computer cartridges)
IBM 3850, *using a honeycomb shell structure.*
Monoclonal antibodies (practical production)
Dr César Milstein, *Head of the Protein and Nucleic Acid Chemistry*

Dr César Milstein

division at the Medical Research Council's Molecular Biology Laboratory, Cambridge, UK, *in conjunction with* Dr Georges Köhler, *German scientist working at the Institute of Immunology in Basle, Switzerland, and* Neils Jerne, *Danish professor.*
Printing ink (ultra-violet)
a research and development team at Coates Brothers Inks *in Kent,* UK.
Submarine pipeline
Ekofisk-Emden gas pipeline, stretching 260 miles under the North Sea.
'Supral' (superplastically-deformable copper-aluminium alloy)
British Aluminium Co., Tube Investments *and* Superform Metals Ltd.
Video games (mass-produced for consumers)
Atari Corp., USA.

1976

AGR (advanced gas-cooled reactor) atomic power station
Hinkley Point, Bridgwater, Somerset, with an output of 1320 MW, based on

a decade of research with the prototype AGR *at Windscale, Cumbria,* UK.
Automatic still camera (microprocessor)
Canon, *Japan – called the* 'AE-1'.
'Charm' (sub-atomic particle)
Stanford Linear Accelerator Center, California, then at the Fermi Laboratory, near Chicago, USA.
Covert Rebus (theory of)
Jack Leslau, *London, who discovered hidden messages in the paintings of Hans Holbein the Younger (fl. 1497–1543).*
D° Meson
Gerson Goldhaber *and* Françoise Pierre *at Stanford University,* USA.
Male contraceptive pill
Professor Salat-Baroux, *Paris, France.*
Marine propeller (tip-vortex-free, TVF)
Gonzalo Perez Gomez and a research team *at Spanish Shipbuilding* SA, *Madrid.*
Mars space probes
NASA's *'Viking 1' and 'Viking 2'.*
'Rapid bursters' (astronomy)
several X-ray satellites, including the 'Copernicus' orbiting observatory and the US *Special Astronomical Satellite,* SAS 3.
Solar-powered boat
Alan T Freeman, *65-year-old former* UK *consultant on railway electrification. The 2.4 metre catamaran 'Solar Craft 1' was powered by two panels containing 50 photovoltaic cells, which drove an electric motor. It proved capable of speeds of 3 mph for two hours on the British Waterways canal at Rugby,* UK.
Super-computer
Seymour R Cray *of Cray Research Inc., Minneapolis, Minnesota,* USA. *With 200,000 integrated circuits to perform 200 million operations per second, it was called the 'Cray 1'.*
Talking card system
Sony, *Tokyo, Japan, to teach tiny*

children language and other kinds of knowledge.
Tandem-aerofoil boat (practical)
Gunther W Jörg, *German engineer. The* TAB VII-5 *'Jörg II' was capable of 135 kmph.*
Unnilseptium
Organesyan *et al.,* USSR.
'US Patent No. 4,000,000'
Robert L Mendenhall, *Las Vegas, Nevada,* USA – *a process for recycling asphalt-aggregate compositions.*

1977
Alkyd paints (fast-drying oil colours)
Messrs Winsor & Newton Ltd, *Whealdstone, North London.*
Artificial pancreas
Dr Wan Jun Tze, *Vancouver, Canada.*
Drug biosynthesis (recombinant DNA techniques)
Genentech, *San Francisco,* USA.
Human-powered aircraft
Dr Paul Macready, US. *Pilot-pedalled by Bryan Allen, the 'Gossamer Condor' flew a figure-of-eight course totalling 1.35 miles (7½ minutes airborne) at Shafter Airport, California to win a £50,000 Kremer Prize.*
Laser-based glass-fibre manufacturing system
Bell Telephone Laboratories, *Murray Hill, New Jersey,* USA, *for harder-than-steel fibre-optic communication lines.*
Neutron bomb
US Military, *based on the work of* Samuel Cohen *in 1958.*
Planet-to-planet unmanned space probes
NASA's *'Voyager I' and 'Voyager II'.*
Pocket television receiver (2 inch screen)
Sinclair Radionics Ltd, UK.
Programmable, polyphonic voice-assignable synthesiser
Sequential Circuits Co.'s *'Prophet 5'.*

1978

Solar power plant
Solar Thermal Test Facility, Sandia Laboratories, *Albuquerque, New Mexico, Sunlight from 222 heliopostats is concentrated on a target 114 ft up in the power tower.*
Space shuttle orbiter
NASA *and* allied aerospace companies *for a re-usable spacecraft programme initiated in 1972. Orbiter No. 101, 'Enterprise,' made a series of approach and landing tests, captive and free, on the back of a specially converted Boeing 747 Jumbo Jet. By 1981, Orbiter No. 102, 'Columbia', made a spaceflight and landed on a runway with astronauts John Young and Bob Crippen on board during a 54½ hour flight.*
Telephonic optical-fibre cables
General Telephone Co. *of California, between Long Beach and a local exchange 5½ miles away in Artesia.*
Visitronic auto-focus (cameras)
Honeywell Inc., US.

1978
ABS anti-skid braking system (motor cars)
Robert Bosch GmbH, *Germany.*
Argon laser (medical use)
Coscas, *France.*
Arneson drive (fixed-shaft, surface-piercing propeller)
Howard Arneson, *Californian inventor, and* Dan Arena, *hydroplane driver, with their prototype offshore powerboat 'Sea Sweep'. Two years later, an adjustable drive was developed and sold to the Borg-Warner Corporation, USA.*
Automatic pressure cooker
Prestige Ltd, *Derby,* UK.
Cyclosporin A (transplant drug)
Dr Tony Allison *in the Clinical Research Centre at Northwick Park Hospital, London, and* Roy Calne, *surgeon at Addenbrooke's Hospital, Cambridge,* UK, *to assist with combined heart and lung transplant operations and combined kidney and pancreas transplants.*
Electronic typewriter
Olivetti *of Italy and* Casio *of Japan.*
Human insulin (recombinant DNA process)
Genentech *of San Francisco,* US, *biotechnology firm. It was produced and marketed by Eli Lilly & Co. as 'Humulin', first in Europe, then in the* USA.
Integrated bodywork bumpers (motor car)
Porsche 928 and 928S, *Germany.*
'Intelligent' typewriter (word processor)
Qyx, USA.
International ultra-violet explorer satellite
to study ultra-violet radiation in outer space.
Laser annealing (silicon chip manufacture)
Stanford University, *California,* US.
Laser for metal removal in aircraft construction
British Aerospace Aircraft Group, *Filton Division, Bristol,* UK.
Magnetic-lift linear-motor car
Japan. It did 310 mph on a test track near Miyazaki.
Plant growth regulators (PGR)
ICI Research Station, *Jealott's Hill, Berkshire,* UK. *It was marketed in 1984 as 'Clipper', 'Cultar', 'Bonzi' etc.*
Pluto's satellite (Charon)
James Walter Christy *at the* US *Naval Observatory at Flagstaff, Arizona.*
Scanning acoustic microscope
Cambridge Instrument Co. Ltd, UK.
64,000-bit memory chip (64K RAM)
IBM, *using their innovation,* SAMOS *(silicon and aluminium metal oxide semi-conductor).*
'Speak and Spell' (instruction machine)
Texas Instruments Inc., *Dallas,* USA.
Store and press telephone call maker
British Telecom.

Test-tube baby
Patrick C Steptoe, *gynaecologist, and*
Dr Robert G Edwards, *physiologist*,
UK. *Louise Brown was conceived in a*
laboratory dish and born two weeks
prematurely by Caesarian section, at
Oldham General Hospital,
Lancashire, UK.

1979
Anti-dive motorcycle
suspension (variable geometry)
simultaneously: Honda's *'Pro Link'*,
Kawasaki's *'Uni Track' and* Suzuki's
'Full Floater', all in Japan.
Artificial blood
Ryochi Naito, *doctor at the*
Fukushima Medical Centre, Japan.
'AI' computer program (to play
backgammon)
Hans Jack Berliner *of Carnegie-*
Mellon University. Called 'Mighty
Bee', it defeated the world
backgammon champion, Luigi Villa,
at Monte Carlo.
Ceramic diesel engine
(prototype)
Cummins Technical Center,
Columbus, Indiana, with the US *Army*
Tank Command.
Computer information-
processing system (Japanese
characters)
IBM. *A 254-key, multiple-shift*
keyboard enabled its operator to call up
more than 7000 Japanese Kanji
ideographs.
Far infra-red laser
Columbia University, *New York*
City, with the Naval Research
Laboratory, *Washington* DC, USA.
Human-powered aircraft
Dr Paul Macready, US. *Pilot-pedalled*
by Bryan Allen, the 'Gossamer
Albatross' crossed the English Channel
– 22 statute miles – in 2 hours 49
minutes.
Human speech machine
IBM *System 370. It is programmed to*
understand a vocabulary of 1000
English words and print out

what it hears.
'Intelligent Pig' (on-line
inspection vehicle for high
pressure natural gas trunk
distribution pipelines)
British Gas.
Masuda wave power barge
Japan Marine Science and
Technology Centre's *'Kaimei,' based*
on Yoshio Masuda's *self-regenerating*
navigational buoys.
Mega-submarine (nuclear-
powered)
USS *'Ohio', 560 ft long and carrying 24*
Trident I missiles with a 4600 mile
range before the need to refuel.
Ricardo comet catalytic unit
Ricardo Consulting Engineers,
Shoreham, Sussex, UK, *with* Johnson
Matthey, *refiners and distributors of*
platinum metals from the Rustenburg
Platinum Mines, South Africa: an
anti-pollution device.
Rubik's cube
Ernö Rubik, *Professor at the*
University of Budapest.
Satellite navigator (miniature)
Thomas Walker & Son, UK, *for*
motor-cruisers and yachts.
Single cell protein process
ICI Agricultural Division,
Billingham, Cleveland, UK; *marketed*
as 'Pruteen', a high protein animal
feedstuff.
Two-megawatt wind turbine
US Department of Energy. *The*
Boeing MOD *1, with blades 68 metres in*
diameter, was installed at Boone,
North Carolina. It was mechanically
unsuccessful.
Ultrasonic scanners (cross-
section views of unborn babies)
based on the pioneer work of Professor
Ian Donald, *Regius Professor of*
Midwifery, University of Glasgow,
UK.
Viewdata systems
Videotex Section of British Telecom
Research Department, *Martlesham*
Heath, Ipswich, UK *– called 'Datel'*
and 'Prestel'.

1980

Sony 'Walkman' – TPS L2

'Walkman' (personal stereo tape-replayer)
Sony, *Tokyo, Japan – called the* 'TPS L2'.

1980
'Balans' chair
Hans Christian Mengshoel, *helicopter pilot, Norway.*
Carbon-fibre tennis racket (injection-moulded)
Dunlop Sports Co. Ltd, *Wakefield, West Yorkshire,* UK.
'Cloned' interferon
Biogen *of Zurich, Switzerland, followed by several other companies. It was produced from genetically engineered bacteria.*
CMOS (complementary metal oxide semiconductor)
Compact disc digital audio system (laser-optically recorded)
Sony, *Tokyo, Japan, with* Philips, *Eindhoven, Netherlands.*
'Dolby C' noise reduction system
Dolby Laboratories, *London,* UK, *and San Francisco,* USA.
Foil propeller (boat propulsion by wave power)
Einer Jakobsen, *Wave Control Co., Norway.*
High quality underwater video camera
Sony Broadcast Ltd, *with the* British

Admiralty's Research Laboratories *at Teddington, Middlesex,* UK.
Holographic cinema (experimental)
N Aebisher *and* C Bainier, *researchers at Franche-Comté University, France, showing the flight of a gull.*
Holographic supermarket terminal
IBM.
Laser cutting of cloth
Hughes Aircraft Corp., *Malibu, California,* USA.
Laser-driven optical system for phototypesetting
Monotype International – *called the 'Lasercomp'.*
Laser repair system (for high-density computer memory chips)
Western Electric Co., USA.
Load-monitored fasteners
Rotabolt Ltd, *Dudley, West Midlands,* UK.
Merchant ship (auto-controlled, rigid, vertically folding sails)
Nippon Kokan, *Japanese shipyard, with the 1600 ton Shin-Aitoku-Maru.*
Microwave thermography
Professor Y Leroy, *University of Science and Technology, Lille, France.*
Personal stereo (weatherproof)
Sony *Tokyo, Japan – called the* 'WM F5'.
Plastic-tipped pen
Berol Ltd *of King's Lynn, Norfolk,* UK, *to improve children's handwriting.*
Quantity-processing computer
Mikroelektronik, *East Germany.*
Rocket boat (liquid-fuelled)
Lee Allen Taylor Jr *of Bellflower, California,* USA. *He was killed when* 'US Discovery II' *disintegrated at high speed.*
Solar maximum mission (satellite)
carrying spectrometers, photometers and imaging devices to monitor the sun

in U/V, X-*ray and Gamma radiation. It was retrieved and repaired by the space shuttle, 'Challenger', in 1984.*

Solar-powered aircraft (for sustained flight)
Dr Paul Macready, *scientist and aviator, California. Piloted by Janice Brown, 'Gossamer Penguin' made a 1.95 mile flight at Edwards Air Force Base in the Mojave Desert, sustaining an altitude of 12 ft for 14 minutes. The following year Macready's 'Solar Challenger' made a 230 mile flight from France to the* UK, *including the Channel crossing. The 'Challenger', powered by 16,000 photovoltaic cells, flew for 5½ hours at a height of 11,000 ft.*

Streamlined cabin tricycle (high-speed)
Allan A Voigt, John Speicher, Steve Wojcik *and* Doug Unkrey *of Versatron Research Inc., Anaheim, California,* USA. *Their Vector tandem reached almost 63 mph through the timed 200 metre traps, whilst their Vector solo reached 59 mph. This was without either stored energy or motor-pacing*

Triple quasar
Steward Observatory *of Arizona University together with the* Smithsonian Astrophysical Laboratory *and the* Bell Telephone Laboratory, *New Jersey,* USA.

Turbocharged motorcycle (production model)
Honda *of Japan – the* 'CX 500 Turbo'.

Waterworks sludge treatment plant
Water Research Centre, *Stevenage, Hertfordshire,* UK, *using a continuous polymer-thickening process. The first full-scale plant went into operation in 1982 for the Anglian Water Authority.*

1981
Air pollution detector (twin laser beam system)
ICI, UK – *the* 'Eagle' *system*.

Anti-interferon
Medical Research Council's Molecular Biology Laboratory *at Cambridge,* UK; *marketed by Celltech, the government-backed* UK *Biotechnology Group.*

First full 32-bit single-chip CPU
Hewlett-Packard, USA – *450,000 individual transistors on a silicon chip measuring ¼ inch square.*

Fully aerodynamic saloon car
The five-cylinder Audi 100. It had a drag factor of 0.30, obtained from such features as a flush windscreen and windows.

Intra-red explorer balloon (Montgolfière type)
launched from Pretoria, South Africa, with four French scientists on board.

Laser-enhanced etching and plating
IBM.

Lightweight plastic engine manifold
Ford Europe *with* Fiberglass Ltd *and* BTR Permali RP, *moulding specialist.*

Micro-chip electric toaster
Russell Hobbs Ltd, *Staffordshire,* UK.

Multi-function bubblejet printing
Canon Inc., *Ota-ku, Tokyo, Japan.*

Nuclear magnetic resonance (NMR) scanner
Thorn-EMI Research Laboratories, *Hayes, and* Nottingham University, UK. *It was installed at the Hammersmith Hospital, West London.*

Simulated listening typewriter (prototype)
Messrs Gould, Conti *and* Hovanyecz *at* IBM, *Yorktown, Virginia,* USA. *It was capable of recognising a limited vocabulary of 5000 words.*

Super-computerised musical instrument
Giuseppe di Giugno, *Italian engineer, for Pierre Boulez, French composer and his 'keyboard' assistant Gerzo, at the Institute of Musical and*

Acoustic Research and Co-ordination, Paris. With the '4X', capable of 200 million operations per second, Boulez composed 'Répons'.
Tandem accelerator mass spectrometer
University of Arizona, *Tucson,* USA, *to carbon-date very small samples of material. The first experiments were carried out by* Paul E Damon, *Professor of Geosciences at the university.*

1982
Abnormal genes (cancer-tumour-causing)
Dr Robert Weinberg *and scientists at the Massachusetts Institute of Technology and* Dr Mariano Barbacid *in the National Cancer Institute, Bethesda, Maryland,* USA.
Artificial gill (sub-aqua breathing device)
Celia *and* Joseph Bonaventura, *Duke University, North Carolina,* USA – *called the 'Hemosponge'.*
Artificial heart (compressed-air-driven)
Dr Robert Jarvik. *The 'Jarvik Mark 7' was implanted in Dr Barney Clark of Wisconsin by William De Vries,* US *surgeon at the Utah Medical Center, Salt Lake City. Clark lived on with the device for 112 days.*
Astronomical observatory (airborne)
NASA, *using a modified* C-141 *jet airplane.*
Automated moulding machines and plating robots (computer-controlled production line)
Oita-Canon Inc., *Ota-ku, Tokyo, Japan, for producing advanced autofocus cameras such as the* 'AF 35 MII'.
Automatic still camera (disc-cassette film)
Kodak, *Rochester,* NY, USA.
Chip for echo cancellation in TV
Philips Research Laboratories, *Eindhoven, Netherlands.*

Colour TV pictures of Venus
taken consecutively through red, blue and green filters, by the descent module of Soviet spacecraft 'Venera 13' (unmanned) which had landed on the planet after a two-year voyage.
Computer-controlled electrical stimulation of paralysed limbs
Dr Jerrold Petrovsky *at the biomedical laboratory of Wright State University, Dayton, Ohio,* USA – *called 'Band Aid'.*
Computer floppy-disc pen
Berol Ltd, *King's Lynn, Norfolk,* UK.
Computer robotic system (RSI)
IBM, *using* AML, *a programming language, for manufacture.*
Denloc tyres
Rothmans-Porsche *'956 C' racing car.*
Domestic robot
Heathkit, *part of the* Zenith Corp., USA – *called 'Hero 1'.*
Electronic still camera
Sony, *Tokyo, Japan, with their Mavica System, using a disc-shaped film which was an electronic video-tape allowing colour pictures to be phoned through to whichever news media required them. While the Sony 'Mavica' had 280,000 picture elements, the* Canon 'ESC', *developed jointly with* Texas Instruments Inc., *had 400,000 picture elements and a charge-coupled device (*CCD*). Canon put their camera on trial at the 1984 Los Angeles Olympic Games.*
'The Fastest Act of Man'
C V Shank *and co-workers at* Bell Telephone Laboratories, New Jersey, USA – *the shortest ever flash of laser light, its pulse lasting only 30 femtoseconds or 0.03 picoseconds.*
Laser gyro inertial navigation system
Honeywell Avionics Division *for the* Boeing 757 *and* 767 *airliners.*
Magnetic bubble binary memory devices (production model)
Texas Instruments Inc., *Dallas,* USA.

Megawatt wind turbine farms
US Department of Energy, *siting three machines with 90 metre-diameter blades at Goldendale Columbia River Gorge, Washington State. They were capable of producing 2.5 MW. A five-turbine farm was also in operation at Byron, California, generating 1.2 MW.*

Monoclonal antibodies (to combat leukemia)
Doctors John Kersey *and* Norma Ramsay *at the University of Minnesota with assistance from immunologist* Tucker LeBien.

Motor car (solar-powered for long distance)
Hans Tholstrup *and* Larry Perkins. *Their 'Solar Trek 1', with a solar-panel roof powering an electric motor through batteries, travelled 4084 km across Southern Australia from Perth to Sydney at an average 23 kmph. Total running time was 172 hours.*

Solar-powered trans-Australian car

Nuclear structure facility (NSF)
UK Government Physics Laboratory, *Daresbury, Cheshire. It included an ion accelerator designed and built for smashing atomic nuclei.*

Organic waste earthworm exploitation process
Dr Clive Edwards *and* researchers *at Rothamsted Experimental Station, Harpenden, Hertfordshire, UK, for use in agricultural and horticultural industries.*

Portable electronic typewriter
Brother Corp., *Japan. Called the* 'EP 20', *it weighed 2 kg and used a 35-dot matrix system.*

Quasar PKS 2000–330
A Wright *and* D Jauncey *at the Parkes Radio Astronomy Observatory, New South Wales, Australia. It is estimated at 13,000 million light years away.*

Rising sector gate (Thames Flood Barrier)
Rendel, Palmer & Tritton, UK *consulting engineers, for the Greater London Council. After two decades of planning and design, the 10 year construction of the barrier was carried out by 26 main building contractors, including two major consortia and several hundred subcontractors.*

Solar power plant
Mojave Desert, California, USA. A 300 ft tower surrounded by 1800 giant movable mirrors generated 10,000 Kw.

Space station (intermediate term)
USSR's *'Salyut 7', with cosmonauts Anatoly Berezovoi and Valentin Lebedev on board, remaining in orbit for 211 days, during which time 87.6 million miles were computed.*

Thermophilic anaerobic effluent treatment plant
researchers at H P Bulmer Ltd, *cider manufacturers, Hereford, UK.*

Unnilenium
G Munzenberg *et al., West Germany.*

'Voice advice' motor car
Peugeot 505 Turbo Injection, France, soon followed by the Renault 11, France and the Austin Maestro, UK.

Word-teaching game (using optical pencil and bar codes)
Texas Instruments Inc. – *called 'Magic Wand'.*

1983
Anti-sound device
University of Leeds *with British Gas, using a computer-controlled sound emitter to cancel out noise.*

'Aquacolor' (true underwater colour ciné)
W Tuckerman Biays *and* Alice de T P Biays, *Islamorada, Florida, USA.*

1983

Aspirated-cylinder-sail catamaran
Commander Jacques Yves Cousteau, France – the 'Moulin-à-Vent II'.
Atomic resolution microscope
National Optical Centre, *Japan.*
Biopol (biodegradable plastic)
ICI Agricultural Division, *Cleveland, UK. After 14 years' research and development, this thermoplastic polymer was produced through a bacterial fermentation process.*
Biosensors
Cambridge Life Sciences *and other companies. Enzymes from living bacteria were linked with silicon chips.*
Carbon-fibre aircraft wing
developed for the UK *government-backed 'Lear Fan 2100' project.*
Colour negative film (extreme speed)
Kodak ASA 1000.
Computer-regulated automatic route-setting and railway-signalling
British Rail, Westinghouse *and* GEC.
Digital scene simulation
John Whitney Jr *and* Gary Demos, *Los Angeles, California,* USA. *Using a Cray X-MP super-computer, the process was first successfully used for 'The Last Starfighter', Lorimar Productions.*
Electronic laser engraving of printing cylinders
Crosfield Electronics Ltd, *Hemel Hempstead,* UK – *called 'Lasergravure'.*
512 K-bit 120 nanosecond (Dynamic Access Memory chip)
IBM, *Essex Junction, Vermont,* USA. *It measured ⅜ inch square.*
Infra-red astronomical satellite (IRAS)
USA *with the* UK *and the* Netherlands. *Discovered or suspected over half a million sources of heat in the universe – including, perhaps, 'Planet 10'.*
Integrated electronic filing system
Canon Inc., *combining microfilm with electronic data bases.*
Intermediate vector bosons (W and Z particles)
Dr Carlo Rubbia, *Italian scientist, and* Dr Simon van der Meer, *Dutch engineer, at* CERN, *the European Centre for Nuclear Research, near Geneva, Switzerland, using a machine that involved the collision of matter and anti-matter beams.*
Joint European Torus (JET)
Experimental Fusion reactor started up at Culham, Oxfordshire, UK.
Laser ranging
Crustal Dynamics *Project at* NASA's *Goddard Space Flight Center, using a reflector-bearing Laser Geodynamics satellite (* LAGEOS*) and a chain of Neodymium –* YAG-*yttrium-aluminium-garnet – lasers.*
MAGLEV (magnetic-levitated linear-propulsion passenger car)
People Mover Group, GEC, Brush Electrical Machines Ltd *and* Balfour Beaty Power Construction – *from Birmingham Airport to the National Exhibition Centre,* UK.
Motor cycle chassis (carbon-fibre)
Armstrong Motor Cycles Ltd, UK.
Newspaper printing plate (double-width plastic)
Mitsubishi, *Japan.*
'Quiteron' (superconducting transistor)
Sadeg M Faris *of* IBM.
RMS (remote manipulator system)
NASA. *A 50 ft long mechanical arm in the payload bay of space shuttle orbiters, for deploying satellites into orbit and collecting them, it was controlled from the flight cabin by astronauts using both a console and TV cameras.*
Robotic manufacture of robots
Yamazaki, *Japan.*
Spacelab
ESA (European Space Agency) *with* NASA.

Tactel
ICI Fibres Division research team –
*an attempt to bring the appearance of
synthetic fibres closer to that of cotton.*
**Tension leg oil production
platform**
Conoco *for the Hutton Field, North
Sea.*
**Tracking and data-relay
satellite (TDRS)**
deployed by NASA's *space shuttle,
'Challenger', the first of four such
satellites.*
**288,000-bit computer-memory
chips (mass-produced)**
IBM.
Videodisc arcade game
'Dragon's Lair'.
**'Voice-advice' still camera
(multilingual options)**
Minolta – *called the* 'AF SV'.
**Winged keel (racing yacht
design)**
Benjamin Lexcen, *marine architect,
Norport Pty Ltd, Australia. It was
used on 'Australia II', which won the
America's Cup.*

1984
All-plastic aircraft
Avtek 400, built in Amarillo, Texas.
**Arc-welding robots (real-time
3-D vision guidance system)**
Oxford University Robotics Group,
Fairey Automation Ltd, GEC *and* BL –
called the 'Fairey MetaTorch'.
Artificial comet
NASA.
Artificially-induced rainfall
Professor A Gagin *and a* team of
Israeli scientists, *by bombarding
clouds with a fine powder of silver
iodine filings.*
**Asbestos safe waste recycling
process**
Dr David Roberts *of King, Taudevin
& Gregson – the 'Vitrifix' process.*
**Automatic still camera (built-in
8-bit microprocessor)**
Canon Inc. – *called the* 'T-70'.

Car compact disc player
simultaneously by Sony *and by*
Fujitsu *in Japan and by* Philips,
Eindhoven, the Netherlands.
Car engine (90 per cent plastic)
Polimotor Research, *New Jersey,*
USA: *tested in a Lola T616 racing car.*
**Cloned antitrypsin gene (body
enzyme)**
National Institute of Health, USA,
Transgene, *France, and* Otago
University, NZ.
Computerised proof-reader
Bert de Pamephilis, US *typographer,
using a standard* ASC II *word-
processing code and voice-synthesiser
board loaded into an* IBM *personal
computer.*
**Computerised translation
machine**
Fujitsu, *Japan. The 'Atlas II' was
able to translate Japanese into English
at 60,000 words an hour.*
**Contact X-ray electron
microscopy**
Thomas J Watson Research Center
of IBM, *New York, in conjunction with
the* National Institute of Health *in
Bethesda, Maryland, the* New
England Deaconess Hospital *and the*
Harvard Medical School, *Boston,*
USA, *by adapting photolithography
normally used for etching micro-
circuits on silicon chips to create
permanent 3-D images of the inside of
living human cells and blood platelets.*
**Genetically-engineered 'Factor
VIII' (haemophiliac blood-
clotting factor)**
21 researchers at Genetech *of San
Francisco and the* Genetics Institute,
Boston, Massachusetts, USA, *with*
Edward Tuddenham *of the
Haemophilia Centre of the Royal Free
Medical School, London.*
**High-speed direct-injection
diesel engine**
Ford Motor Co., *Europe, as used in
their Transit van.*
Holocamera
Newport Corporation, *Fountain*

1984

Valley, California, USA.

Holocopier (rapid hologram copying machine)
Applied Holographics, Essex, UK. It produces over 400 copies per hour.

Hologram-illustrated magazine cover
'The National Geographic', Washington DC, USA.

Holographic credit-card
VISA, soon followed by banks throughout the world.

Hubot (completely integrated robot)
Hubotics Co., Carlsbad, California, USA – a TV Screen/computer/FM stereo receiver.

Human-powered hydrofoil (catapult launch)
Alec Brookes and Allan Abbott of Pasadena, California, USA. The 'Flying Fish' was successfully pedalled for 2 km in 6½ minutes by Steve Hegg.

Kodavision (amateur video film)
Kodak, Rochester, New York, USA.

Laser gravity-wave detector
Ronald Drever and researchers at the California Institute of Technology, Pasadena, USA.

Laser light compressor
Jean-Marc Halbout, Daniel Grischkowsky, Anne C Balant and Hiroki Nakatsuka, scientists at IBM's Thomas J Watson Research Center, Yorktown Heights, New York, to generate 12-femtosecond, ultra-short light pulses – at the rate of 800 per second.

'LIDAR' (transportable laser radar)
researchers at Lincoln Laboratory, Massachusetts Institute of Technology, USA.

'LIF' (laser-induced fluorescence)
Richard Neil Zare, laser chemist at Stanford University, California, USA, to detect chemicals in gases and liquids.

MMU (manned manoeuvring units)
NASA, enabling space shuttle astronauts, McCandles and Stewart, to make the first-ever untethered spacewalk 110 yards from 'Challenger'.

1024K-bit integrated circuit
simultaneously by Hitachi, NEC, NTT Atsugi Electrical Communications and Toshiba, all of Japan.

1,048,576-bit computer memory chip
IBM.

Parking meter (credit-card operated)
GEC Traffic Automation Ltd, Borehamwood, UK.

Pedalectric tri-cyclecar
Sinclair Vehicles Ltd of Coventry, UK, in conjunction with Lotus Cars of Norfolk, UK – called the 'Sinclair C5'.

Pocket television (flat screen)
simultaneously by Sony, Seiko and Casio of Japan. A wristwatch television followed soon after.

Portable compact disc player
Sony, Tokyo, Japan – called the 'Sony D50'.

Robotic electronic component assembler (high-speed)
ASEA, Sweden — the 'IRB 1000'.

RTM III (rock-trenching machine)
remotely controlled submarine robot to cut two trenches across the sea-bed of the English Channel for the 2000 MW interconnection between both the Central Electricity Generating Board and Electricité de France power station networks.

64K-bit 20-nanosecond dynamic access memory chip
Stanley Schuster and researchers at IBM's Thomas J Watson Research Center, Yorktown Heights, New York, USA. It is capable of discharging sixteen bits of information at a time, using a four-device memory 'cell'.

Spring and damper (plastic and rubber)
National Engineering Laboratory, Glasgow, UK.

Supersonic forward swept-wing aircraft
US *Grumman* X-29, *a single-seater fighter-plane with wing-joints using Mylar and carbon-fibre. First tests were held at Edwards Air Force Base, California,* USA.

'Top Quark' (sub-atomic particle)
Carlo Rubbia *et al.,* CERN, *Geneva, Switzerland.*

Unniloktium
G Munzenburg *et al.,* West Germany *and the* Dubna Research Institute, *Moscow.*

1985
Jewish Year 5746

1985
ALERT (automatic logging of events and ruptures in tyres)
Lartrack, UK, *with* Honeywell, US. *A sensor is fitted between each tyre and its rim.*

Anti-ICBM/Spy satellite high-energy ground-baser laser
by USA *and* USSR *separately.*

'Anxiety chemical' (human brain)
Alessandro Guidotti *and* Erminio Costa, *National Institute of Mental Health, Maryland,* USA.

Artificial tooth root
Furukawa Electric Company *with the* Dentistry Department of Nihon University, *Japan. Made of 'shape memory' nickel-titanium alloy.*

Atmosphere trace molecule spectroscope (ATMOS)
Honeywell *for use by* NASA's *Jet Propulsion Laboratory in Pasadena, California,* USA. *An infrared interferometer installed on Spacelab 3 and pointing towards the sun to investigate the chemicals which affect the ozone layer.*

Captopril (orally active angiotensin-converting enzyme-inhibitor)
E R Squibb & Sons, *Princeton, New Jersey,* USA. *Based on over a decade of research by* Messrs Cushman *and* Ondetti *at the Squibb Institute for Medical Research. This* ACE *inhibitor was marketed as 'Capoten'. It was followed by 'Enalapril' as developed by researchers at Merck, Sharp and Dohme, West Point, Pennsylvania,* USA.

Car body (self-coloured, injection-moulded polypropylene)
Austin Rover, *Coventry,* UK. *As incorporated in the 'MG* EX-E*'.*

Catamaran powerboat (transatlantic)
Cougar Marine, *Hampshire* UK. *The 65 ft Virgin Atlantic Challenger. After completing 2973 nautical miles at an average 41.5 knots, the Challenger hit submerged debris and sank less than 200 miles from the* UK *coast.*

CD-ROM (compact-disc read-only memory)
Hitachi, *Japan under licence to Philips, Eindhoven, Netherlands. Their '1502S', with a parallel interface for the* IBM-PC*. By storing 522 Megabytes, 270,000 A4 pages can be placed on a single* CD*.*

Cellular radiophone (nationwide network)
Racal-Vodaphone Ltd, *Newbury, Berkshire,* UK. *Also 'Cellnet' by* British Telecom *and* Securicor.

Cloned leprosy genes (for vaccines)
Ron Davis *et al., Stanford University,* USA.

Cockpit safety capsule (circuit catamaran powerboats)
Chris Hodges, *boatbuilder, Norfolk,* UK, *following the death of his partner Tom Percival in a race. First saved a life when Bob Spalding,* UK, *somersaulted his catamaran at the Sacramento Grand Prix,* USA.

Comet-intercepting spacecraft
Space & Communications Division of British Aerospace Dynamics

'Giotto': comet-intercepting space probe

Group *leading a component consortium from 9 different countries for the European Space Agency. Called 'Giotto', it was designed to chemically analyse and also take 'close-up' photographs of Halley's Comet (from 300 miles) before being destroyed by it.*

Compact disc automobile-navigation system
Philips, *Eindhoven, Netherlands – called 'Carin', a compact-disc-personal-computer for maps, information and music.*

Computer data wristwatch terminal (using microsoft BASIC)
Seiko, *Japan. Called the* 'RC 1000'; *80 'pages' of text transmittable from any computer via an* RS-232C *interface. Includes the time in up to 80 cities around the world.*

Computer-generated, synthetic text-to-speech system (intonation and voice-range options)
Dr Dennis H Klatt, *computer scientist at Massachusetts Institute of Technology,* USA, *in conjunction with* Ed Bruckert *and* Walt Tetschner *of the Digital Equipment Corporation, Maynard, Massachusetts,* USA. *Called 'DECtalk' it is able to generate exact pronunciations for over 20,000 words.*

Computerised submarine-hunting detection device
Gresham-CAP, *New Malden, Surrey,* UK *for the Royal Navy. Called*

'SEPADS' = *Sonar Environmental Prediction and Display System.*
Cryostatically-immersed traction magnets
Bangor *and* Oxford Universities, UK, *using liquid helium.*

Domesday project
Teachers, schoolchildren et al., throughout the UK. *Two videodiscs storing both digital data and moving and still pictures concerning the* UK *during this year. 'Re-read' by combining a Philips LaserVision player,* BBC *micro-computer with* VDU *and cursor.*

Driverless six-wheeled automobile (experimental)
Researchers at the Robotics Institute, *Carnegie-Mellon University, Pittsburgh,* USA. *Using stereo* TV *cameras, acoustic soundings and computer memory bank.*

Electricity-conducting polymer
Terje Skotheim *and* researchers at *Brookhaven National Laboratory, New York. Called the* 'PP-CoPc'.

Filmless X-ray system
Koyo *of Japan.*

Gallium-arsenide, 6000-transistor gate array chips (mass-produced)
Honeywell *at their factory in Richardson, Texas,* USA, *under contract to the* US *Rockwell Corporation for use in aerospace and defence.*

Geodesic-domed house (triple-skin ventilation roof system)
Nectar Domes Ltd, *Purley, Surrey,* UK.

Image digitiser
Optronics, *Cambridge,* UK. *It converts an image on a page to a digital electronic signal that can be printed out or transmitted over vast distances via computers, fibre-optic data transmission lines and satellites.*

Integrated circuits (half-micron line width manufacture)
IBM scientists *at Thomas J Watson Research Center, Yorktown Heights, New York* USA.

Largest high-definition television screen in the world
Sony, *Tokyo, Japan. Called the 'Jumbotron' and measuring 40 metres wide by 25 metres high, it was demonstrated at Expo '85, Tsukuba, Japan.*

Laser coronary angioplasty
James Forrester, Warren Grundfest *and* Frank Litvak, *physicians at the Cedars-Sinai Medical Center, Los Angeles, California,* USA, *using a 'cool' Xenon chloride laser.*

Laser-disc electronic filing system
Philips, *Eindhoven, Netherlands – called 'Megadoc', with screen or facsimile-printout retrieval options.*

Laser-optical storage system
Jerome Drexler, *Drexler Technology Corp.,* USA – *called the 'Lasercard'.*

Lean-burn automobile engine (experimental)
research engineers, Ford Motor Company, *Dunton, Essex,* UK – *using variable inlet geometry.*

Linear-motor car
Japan Airlines.

Miniature radio (3 mm thick)
Sony, *Tokyo, Japan – called the* 'ICR-101', *nicknamed the 'Credit Card Radio'.*

Mission adaptive wing
Boeing FI-11 *as modified in conjunction with* NASA *and the* USAF. *Tested at Edwards Air Force Base, California,* USA.

Motor yacht (entirely moulded in cord carbon-fibre)
147 ft 'Sterling Lady'. Jon Bannenberg, *London yacht designer,* Sterling Yacht and Shipbuilders *of Miami, Florida,* USA, *and* Nishii Shipyards, *Japan.*

Oil spill solidification process
BP Research Centre, *Sunbury-on-Thames,* UK, *after 5 years' field research.*

Omnigripper (robotic hand device)
Centre for Robotics, *Imperial College, London,* UK.

Positron emission tomography
Michael Phelps *et al, University of California Medical School, Los Angeles,* USA. *To predict, by scanning, likely sufferers of Huntington's Chorea.*

Proton-antiproton collider (synchroton energy equivalence of 400,000 GeV)
CERN, *Switzerland.*

Scanning tunnelling microscope
simultaneously by Calvin F Quate *and researchers at Stanford University,* USA *and* Gerd Binnig, Heinrich Rohrer *and researchers at* IBM *Research Laboratory, Zurich, Switzerland. Able to resolve features one-hundredth the size of an atom.*

Soft bifocal contact lenses
Softsite Contact Lens Laboratory, *Coral Gables, Florida,* USA, *following 7 years' research and 20,000 experimental lenses.*

Solar-powered pay-phone
Plessey Telecommunications Ltd, UK, *for use in mountain chalets and developing countries.*

Spaceflight-proof maser atomic clock
Hughes Aircraft Co., *Malibu, California, under a 3-year contract to the* US *Naval Research Laboratory, Washington* DC. *The clock is accurate to within 1 second in 30 million years.*

'Spectra 900' (synthetic ultra-strong fibre)
Allied Corp., *New Jersey,* USA. *It is ten times stronger than steel, but light enough to float.*

Super-scanning electron microscope (computer-controlled)
Professor Albert Crewe, *Enrico Fermi Institute, Chicago University,* USA. *8 ft tall, weighing 1 ton and able to resolve detail down to half of one angstrom.*

Telescope (for millimetre and sub-millimetre wavelengths)
Rutherford Appleton Laboratories, UK, *in conjunction with* Genius

Company, *Ijmuiden, Netherlands. Installed near the summit of Maunea Kea, Hawaii – remote computer-controlled at the Royal Observatory, Edinburgh,* UK.

Thermal electronic typewriter
IBM, *Lexington, Kentucky,* USA. *Called the 'Thermotronic I'.*

3-D television
Matsushita Electric Industrial Co., *Japan, using a 14 inch screen and requiring no special glasses.*

Three-rotor rotary car engine (experimental)
Mazda, *Japan. As installed in the 'MX-03', also fitted with four-wheel steering.*

Ultra-fast Gallium-arsenide micro-chip
Honeywell Physical Sciences Center, *Bloomingdale, Minnesota,* USA. *Enabling the electrical signal to travel from the point of input to output point in just over 11 trillionths of a second.*

Ultrasonic bubble-cleansing domestic washing machine
Japan Ace, based on research at Nihon University – detergentless.

Unducted fan aero-engine
General Electric Aero-Engine Co., *Cincinatti, Ohio,* USA. *Principle also under evaluation by engine manufacturers in Europe and the* USSR.

Video 8 camera
Sony, *Tokyo, Japan. It weighs less than 5 lb.*

Proposed for 1986

Airbus A320 (civil airliner).

Electra 225
British Rail; *Inter-City streamlined locomotive, electrically-powered, capable of 225 kmph (140 mph).*

'Hipparchus' (spaceprobe)
European Space Agency – *to plot precisely the accurate position of 100,000 stars.*

International solar polar mission
NASA – *'Ulysses'.*

Jupiter space probe
'Galileo'.

Oil-production vessel (dynamically positioned)
BP Petroleum Development Ltd, UK *– directly connected to sub-sea well-heads.*

Shipbuilding arc-welding robot
Kemppi, Rosenew *and* Wartsila Shipyards, *Japan.*

Space telescope (called 'The Hubble')
NASA, *fitted into orbit by the space shuttle and used outside the earth's atmosphere to observe planetary systems as charted by 'Hipparchus'.*

Variable-geometry vertical-axis wind turbine
Dr Peter Musgrove *of Reading University,* UK, *and built by a consortium headed by McAlpine & Sons at Carmarthen Bay, Wales.*

Notwithstanding a 'nuclear winter',
the 'greenhouse effect', an alien
invasion and other, unforeseen,
cataclysms, the following innovations
might be expected during the next 50
years:

Adiabatic engine (production
model).
Air-phone satellite system.
Artificially intelligent, fully robotic
domestic servants and/or
companions.

Biological computer.
Biological pesticides (mass-
produced).

Cellular holophone.
Circular and revolving floating
restaurant.
Civil airliner (vtol, composite plastic
construction and hydrogen-
fuelled).
Cloned superhumans, superlivestock
and supervegetables.
Computer holographics.
Constellation of laser battle stations
in space.

Discovery of the cellular causes of
certain types of cancer.
Discovery of new elements and new
sub-atomic particles.
Discovery of Planet 10 in our solar
system.
Cryogenic deep-freezer (domestic
production model).

Drugs, administered via nasal and
specifically sited transdermal
routes.

Earth-orbiting hotel and planetary
cruise-liner (laser-propelled).
Electric-engined, plastic-bodied
saloon car (long-range and solar/
light rechargeable).
'Electronic secretary' (production
model).
Electronically-stored and displayed
sea charts.
Energy ultra-efficient, pollution-
free, 'domed-in' conurbations.
England-France Channel Tunnel.
Eradication of acquired immune
deficiency syndrome (AIDS).
Exchangeable cassette-array traction
batteries.

Fibre-optic world communication
network.
Fifth-generation artificially
intelligent computerised personal
adviser (expert in medical/legal/
accounting/estate agency/do-it-
yourself/educational matters).
Fleets of 'commercial sail' merchant
ships (steam/electric/magnetic/
propulsion options).
Fleets of freight-carrying airships
and chains of airship mooring
masts.
Fleets of OTECS (oceanic thermal-
energy collectors).

Fleets of 'space tugs' (orbital manoeuvring vehicles), to journey between shuttles, spacelabs and lunar/planetary bases for retrieval and delivery work.

400-mph powerboat.

400-passenger, computer-controlled, five-masted motor-sailing barque.

Genetically engineered disease-and-pest-resistant crops.

Geothermal-energy power stations (practical).

Giant artificial oasis (irrigation technology).

Hologramophone (compact discs).

Holovideo-polyphonic domestic games.

HOTOL re-usable spaceship (horizontal take-off and landing).

Hydrogen factory (using chloroplast technology).

Iceberg-irrigation systems.

Interstellar precursor.

Ion-drive-engined projectile (100,000 mph capacity).

Inference engine.

Knowledge-based processors.

Laser-equipped weather satellites.

Long-range manned spacecraft.

Lunar survey and mining station (including hotel).

Manned planetary landing (other than lunar).

Motor car (production model): to include constantly variable transmission, computer-controlled 'voice-advice' navigation systems, four-wheel steering, ID-car operation and adjustment, plastic-injected tyres in matching colours and voice-command of dashboard instrumentation.

Nationwide chain of lead-free petrol stations.

Nationwide chain of nuclear-fusion power stations.

Natural tooth enamel (synthetically produced).

Nuclear-pumped, X-ray laser gun (star wars).

Offshore metropolis.

Offshore wind-turbine platforms.

Permanently manned sea-bed laboratories.

Omni-purpose international credit/identification micro-chip (implanted subcutaneously).

'Particle beam' weapon (star wars).

Permanently manned space station factory: to produce drugs, super-glass, ultra-strong cast-iron, fibre-optic materials and gallium-arsenide crystals; powered by solar energy.

Public lending library stocking computer programs, videodiscs, compact discs, videocassettes and holographically illustrated books.

Self-service banks (rapid deposit and encashment).

SETI probes (search for extra-terrestrial intelligences).

Ships and submarines (propeller-less and rudderless): using superconducting electro-magnetic thrusters, on both hull and sea-bed, to attain 100-knot speeds.

700-passenger airliner.

'Smart pills' (total control of the precise site and rate of release).

Solar heliospheric observational spacecraft.

Supersonic land speed record car.

'Talking typewriter' (production model): able to read back what is typed into it.

Three-masted cutter-rigged yacht (112 ft length overall): requiring only one man to operate the computer-controlled automated

rigging and electronic navigational
equipment.
Tidal power stations.
Time machine (long-range).
200-miles-per-gallon motor car.

Ultra-violet explorer satellite.
Unducted-fan aero-engine (contra-
rotating, multiple-layer
propellers).

Video-discs and compact discs
(erasable).
Voice-operated, personal-typefaced,
proportional-spaced word-
processor.

Wind-rain-sun-sensitive energy-
versatile roof tiles.

*(Notification to the publisher when
these, or other, ideas are realised would
be most welcome.)*

Annual Register of World Events, The, (from 1945), London, published annually

BRYDSON, JA, *Plastics Materials*, London, 1982

HAYDN, JOSEPH and VINCENT, BENJAMIN, *Haydn's Dictionary of Dates and Universal Information relating to all Ages and Nations*, London, 1881

Innovation and Achievement: Scotland's Place on the World Map (Scottish Development Agency), Edinburgh, 1984

FRANZBERG, MELVIN and PURSELL, CARROLL W, JR, (editors) *Technology in Western Civilisation*, Vols I and II, Oxford, 1966/1967

LARSEN, EGON, *A History of Invention*, London, revised edition 1971

NEEDHAM, JOSEPH with WANG LING, *Science and Civilisation in Ancient China*, Vols I–V, Cambridge, 1956–1980.

Queen's Annual Award to Industry, The, published in the London Gazette, London, 1966–1984

READERS' DIGEST ASSOCIATION, *The Inventions that Changed the World*, London, 1982

ROBERTSON, PATRICK, *The Shell Book of Firsts*, London, 1974

Smithsonian Book of Invention, The, Washington DC, 1978

Bibliography

Entries in this index are followed by the year in which they appear, not by page numbers. Entries falling in the same year are listed alphabetically.

Index

aircraft carrier
    adapted ship 1911
    custom-built 1922
    nuclear-powered 1960
    super 1955
aircraft detection, electromagnetic 1934
aircraft engine
    back-pack 1961
    rotary 1887, 1907
    unducted fan 1985
aircraft radio-location 1935
aircraft refuelling, in-flight 1923
aircraft undercarriage
    hydraulically operated retractable 1929
    pneumatic-tyred 1906
    retractable 1911
    wheeled 1906
aircraft wing
    carbon-fibre 1983
    mission-adaptive 1985
    slotted 1921
airflow ventilation 1964
airfoilcraft 1970
airliner
    Airbus 320 1986
    automatic landing 1965
    jet-powered 1949
    jumbo jet 1969
    pressurised passenger cabin 1938
    supersonic 1968
    turboprop 1948
airscrew, variable-pitch 1924
airship
    diesel-engined 1929
    non-rigid/helium-filled 1918
    rigid 1900
    rigid helium-filled 1925
    steam-powered 1852
airship hangars 1916
airship-tethering mast 1911
'airstrip' (medical dressing) 1957
alarm-clock tea-maker 1902
alcohol still 1832
ALERT (tyre sensor) 1985
ALGOL (computer language) 1958
alizarine red 1831
alkalimeter 1815
Alka-Seltzer 1931
alkyd paints 1977
allopurinol 1962
alpaca 1848
alphabet
    Arabic 512
    Aramaic/Assyrian 720 BC
    Aryan 300 BC
    Chinese 1880
    complete 1300 BC
    Ionian-Greek 400 BC
    Phoenician (22-letter) 1100 BC (30-letter)
        1500 BC
    Sinaitic 1600 BC
    Slavonic 862
alphabetic script, phonetic 775 BC
Alphonsine tables 1253
ALSEP (experiment package) 1969
alternation of generations 1819
alternator
    electric 1878
    high frequency (Alexander) 1906
aluminium 1827
    chemically produced 1854
    electrolytically produced 1886

ambulance 1792
americium 1944
amines 1849
ammonia 1774
    synthesis process 1909
ammonia-soda reaction 1810
ammonium picrate 1885
amniocentesis 1956
amphibious vehicle 1805
    DUKW 1942
    petrol-engined 1905
amphicar 1962
amplifier, rock music 1949
amplitude modulation 1915
amputation 1679
amyl cinnamic aldehyde 1926
anchor
    C.Q.R. 1933
    stockless 1821
    Trotman's 1853
anchor escapement 1680
anchor fluke 592 BC
androsterone 1931
anaesthesia, ether 1842
anaesthetic
    chloroform 1847
    ether 1846
    intravenous 1902
    local 1884
    nitrous oxide 1844
    novocaine 1904
    spinal 1855
Anémocorde 1789
anemometer 1709
    cup 1846
    pressure 1775
aniline 1826
aniline dye 1856
aniline yellow dye 1858
animal cells 1839
animal husbandry 11000 BC
animalcules (bacteria) 1683
Anno Domini dating system 532
anorthosite 1971
anthropoglossus 1864
antibiotic drug 1943
antibodies
    chemical molecular structure of 1969
    monoclonal 1975, 1982
anti-cyclone 1863
anti-depressant drug 1953
anti-freeze 1927
anti-helium 3 1972
anti-interferon 1981
anti-knock petrol compound 1921
anti-matter particles 1972
antimony 1490
anti-proton 1955
anti-sound device 1983
anti-tank grenade projector 1941
anti-tank gun 1917
antitrypsin gene, cloned 1984
anti-Xi-zero (particle) 1963
'anxiety chemical' (human brain) 1985
Apollo experiment package 1969
Apollonicon (barrel-organ) 1817
apple, Granny Smith 1868
aquacolor (cine) 1983
aqueduct 700 BC, 312 BC
aqualung 1943
arc furnace, electric 1894
arc lighting 1879

Disprin 1948
dish-washing machine 1889
dividing engine 1740, 1768
dividing machine, circular 1872
diving bell 1531, 1717
diving dress 1715
   closed 1837
   open 1819
DNA 1953
   biosynthesis 1956
DNA, recombinant 1973
Döbereiner lamp 1823
Dolby noise reduction systems 1966, 1970, 1980
doll
   talking 1823
   toy 1413
doll's house 1558
dome
   design 1432
   iron 1811
   single-shell 532
Domesday project 1985
domesticated animals 11000 BC
   dogs 7700 BC
   goats 8050 BC
   pigs and cattle 7000 BC
   sheep 7200 BC
domesticated crops 9000–8000 BC
   peas and beans 7000 BC
Donati's comet 18;8
door sockets, stone 4500 BC
doors, Baptistry 1452
Döppler effect 1851
   in canal rays 1909
double-decker bus, enclosed 1909
drainage engine 1623
draw-loom 100 BC
drawing pen, tubular-nibbed 1954
dredger 1561
drill
   dental see dental drill
   electric hand 1895
   stone-working 2500 BC
drilling, compressed-air 1857
drive-in bank 1937
driving belt, rubber 1825
driving chains 1864
drosophila 1910
drug biosynthesis 1977
Drummond light 1826
dry cleaning, clothes 1855
Duffy blood group system 1950
DUKW (amphibian vehicle) 1942
dumet 1913
Dunlopillo (foam rubber) 1929
duplex burner 1865
duplicating machine
   hectographic 1923
   thermographic 1973
duralmin 1910
dust filter 1906
dustbin 1883
dust-jacket, book 1833
dye-coupler colour process 1914
'Dyform' strand (prestressing) 1956, 1965
dynamic memory cell 1967
dynamite 1867
dynamo, self-acting 1866
dynamometer 1750
   rolling-road testbed 1904
dynastic divisions 250 BC
dysprosium 1886

E

Earth Resources Technology Satellite 1972
earthquake alarm 132
earthquake measurement scale
   Mercalli 1902
   Richter 1935
eagle, mechanical 1245
ear, tube structure of 520 BC
eau de Javel 1787
eau Grison (fungicide) 1852
ebonite 1851
echo-sounder 1917
   transistorised 1958
EDSAC (calculator) 1949
EDVAC (computer) 1949
EEG 1929
effluent treatment plant, thermophilic
   anaerobic 1982
eidograph 1821
eidophusikon 1781
Eiffel Tower 1889
einsteinium 1953
ejection seat 1938
elastic web 1830
Elastoplast (bandage) 1928
Electra 225 (locomotive) 1986
electrets 1919
electric-arc lamp 1808
'electric aura' 1939
electric battery 1770
electric bell 1831
electric candle 1876
electric car 1874
electric chair 1890
electric currents, secondary 1838
electric-generator windmill 1890, 1931
electric light 1834
   for towns 1880
electric machine
   frictional 1660
   Leyden-jar-powered 1855
electric motor 1829, 1837
   direct-current 1873
   induction 1888
   linear 1985
   miniature 1880
   synchronous 1902
electric telegraph 1837
electric telegraph key 1833
electrical properties (amber) 600 BC
electrical resistance 1827
electricity, galvanic 1790
electrocardiogram 1887
electrocardiograph 1903
electro-convulsive therapy 1937
electro-encephalogram (EEG) 1929
Electrolux refrigeration system 1922
electromagnet 1824
electromagnetic balance 1831
electromagnetic induction 1831
electromagnetic manufacture 1854
electromagnetic motor 1829
electromagnetic radiation 1873
electromagnetic wave transmission 1864
electromagnetic waves 1887
electron 1896, 1952
electron accelerator, Cambridge 1962
electron bombardment engine 1960
electron diffraction by crystals 1926
electron microscope 1929
electronic filing systems, integrated 1983

harvester 1858
hat
  Bowler 1849
  high beaver 1797
  mechanically-tipping 1896
hatchback estate car 1954, 1961
hearing aid
  electric 1901
  electronic 1935
  transistorised 1952
  valve system 1923
heart-lung machine 1953
heart pacemaker internal 1957
heart valve, artificial 1952
heat-bottled food 1809
heat theorem 1906
heater, electric 1887
heavy water 1940
heckelphone 1904
hectographic duplicating 1923
helicopter
  contra-rotating rotors 1936
  gas-turbine-engined 1951
  jet-prop 1953
  jumbo 1968
  main and small tail rotors 1939
  petrol-engined 1907
  production model 1940
  steam-engined 1870
helicopter model 1784
helicopter top 320
'heliographie' 1816
heliotrope (surveying instrument) 1821
heliotropin 1879
helium 1894
  in sun's chromosphere 1868
  liquid 1908
herbicide
  2, 4-D 1941
  MCPA 1941
hemp-spinning machine 1810
hexameter 190 BC
hexameter machine 1846
hieroglyphic texts 2600 BC, 2400 BC
High-Speed Train (HST) 1975
'Hipparchus' (space probe) 1986
histidine 1896
'History of the World' 1311
hobby horse 1790, 1817
hodometer 1528
hodometer-compass 1596
hoe blade, stone 5000 BC
hoist, hydraulic 1912
'Holland Circle' (circumferentor) 1612
holmium 1879
holocamera 1984
holocopier 1984
holograms 1947
  uses of 1968, 1980, 1984
holographic storage material 1968
homeopathy 1810
hormones 1902
  human growth (HGH) 1956
  sex 1927
  see also insulin; thyroxine
horn
  bulb 1871
  electric motor 1905
horse, artificial 1673
horse-bit 1500 BC, 1200 BC
horse-box 1836
horse-shoes 890
hose-pipe, fire-fighting 1672

hosiery factory, children's 1838
hosiery-knitting machine 1589
hospital, public 372
'hot bulb' engine 1890
'hounds and jackals' (game) 1700 BC
hourglass set 1475
hovercraft 1955
  flexible skirt 1961
  full-sized 1959
  passenger 1962
  production model 1964
  tracked 1962
'Hubble' (telescope) 1986
Hubot (robot) 1984
'hula-hoop' (game) 1958
human anatomy book 1543
human speech machine 1979
humidity control 1906
Huntington's Chorea 1985
hydrapsis 1690
hydraulic crane 1965
hydraulic press 1648, 1795
hydraulis (organ) 250 BC
hydrazine 1887
hydro-aeronaut 1810
hydro-aeroplane 1910
hydrochloric acid 1648
hydrocycle
  aerial propeller 1913
  triplet 1895
hydro-electric power 1867, 1881
hydrofoil
  air-cushion swing-wing 1904
  amphibious gas-turbine-engined 1959
  boat 1861
  commercial passenger 1953
  high-speed 1919
  human-powered 1984
  passenger waterjet-propelled 1974
  rotary foil 1876
  steam-powered 1897
  turbojet-engined 1952
hydrofoil ladder 1905
hydrogen 1766
  heavy isotope 1940
  see also deuterium
hydrogen bomb 1952
hydrogen peroxide 1818
hydrogenation of oils, catalytic 1899
hydrokimeter 1874
hydroplane 1872
  aerially propelled 1907
  multi-step/petrol-engined 1908
hydropneumatic suspension 1955
hydroponics 1929
hydrotherapy, revival 1797
hydrostatic bed 1832
hydrostatics 220 BC
hydroxy citronellal 1905
Hygeiopolis 1876
hygrometer 1664
  ether 1820
  hair 1781
hymnbook, Czech 1501
hyperbola 220 BC
hypodermic syringe 1869
hyposulphite of soda 1842

I

'I Speak Your Weight' machine 1928
'IBM System 360' 1964

five-valve 1932
FM 1952
motor car reversing light 1921
motor car safety belt 1959
motor car self-starter 1911
motor car speedometer 1901
motor car tyres
  cord-type 1900
  Denloc 1982
  logging sensor 1985
  pneumatic 1895
  radial-bodied 1937
  radial-ply 1953
  snow-chains 1904
  tubeless 1947
motor car wheels, detachable wire 1905
motor car ventilation, air-flow system 1964
motor car windows,
  electric lift 1946
  rear-window heating 1948
motor car windscreen
  curved 1914
  toughened float-glass 1959
  wipers 1916, 1917, 1920
  zone-toughened 1961
motor caravan 1902
motor carriage, coal gas 1863
motor crash helmet 1902
motor-cruiser, glass-fibre 1951
motor-cycle
  anti-dive suspension 1979
  carbon-fibre chassis 1983
  centre-frame engine 1901
  desmodromic valves 1956
  electric self-starter 1913
  four-cylinder 1895
  in-line, air-cooled, four-cylinder engine 1903
  parallel vertical twin-engine 1904
  petrol-engined 1885
  pneumatic-tyred 1894
  quantity-produced 1895
  rotary-engined 1892, 1974
  telescopic fork 1904
  turbocharged 1980
  V-Eight 1908
  V-twin two-stroke engine 1902
  Wankel-engined 1974
motor horn, electric 1905, 1906
  tune-playing 1929
motor-racing circuit, banked 1907
motor-scooter 1915
  snow 1959
motor speech centre (brain) 1861
motor vehicles, power steering 1926
Motorola (radio) 1929
motorway 1924
Moulton bicycle 1960
mouse-trap, spring-operated 1910
mouth organ 1821
mower, circular-bladed 1805
mowing machine 1810
  see also lawn mower
mule (spinning machine), high-speed self-
  acting 1824
'multics' (computer system) 1965
multiplane, Venetian-blind 1907
multiplication sign 1631
mummy 2400 BC
muon neutrino 1962
mural arc 1688
mural circle 1789

music
  computerised algorithmic 1961
  printed 1472
music synthesiser
  electronic 1954
  voltage control 1964
musical instrument, super-computerised 1981
musical notation 1300 BC, 1026
  rhythmic 1200
musk xyol and ambrette 1891
musket, breech-loading 1786
mustard gas 1917
mustard mill 1845
mustard powder 1850
myographion 1850

N

nail violin 1770
napalm (incendiary) 1942
naphthalene 1819
narrow boat
  motor 1912
  steam-powered 1864
natural gas, North Sea 1965
navigation system, laser gyro inertial 1982
needle gun 1837
neodymium 1885
neon 1898
neon lighting 1910
neoprene 1932
nephoscope 1868
Neptune (planet) 1846
neptunium 1940
'Neracar' 1921
nerve
  depressor 1866
  optic 1538
nerve gases 1944
nervous reflex 1833
net-weaving machine 1830
neutrino 1931
Neutrodyne receiver 1919
neutron 1932
neutron bomb 1977
'New Science, The' 1537
newspaper 1609
  daily 1650
newspaper printing plate 1983
niacin 1937
nichrome 1906
nickel 1751
  process for pure 1890
nickel-cadmium storage cells 1908
nickel-ion alkaline storage battery 1908
nickel-plating 1869
Nike-Zeus (missile) 1959
niobium 1801
Nipkow scanning disc 1884
'nipple'-gun, grease-filled 1919
Nissen hut 1916
Nitonol memory metal 1960
nitric acid 1287
nitro-cellulose 1833
nitrogen 1772
nitrogen-from-air process 1903
nitroglycerine 1847, 1864
nitrous oxide anaesthetic 1844
Nivea cream 1911
nobelium 1958
nocturne (musical piece) 1834

sail
  and rigging 100 BC
  Genoa jib 1927
  spinnaker 1865
  windmill-spring 1772
sail material
  nylon 1945
  rayon 1937
  terylene 1952
sailboard
  commercial 1970
  fibreglass 1964
  prototype wooden 1958
sailing dinghy, polypropylene 1973
salad cream 1927
saloon, swinging 1874
saloon car 1903
Saltaire (model village) 1853
Salvarsan 1909
samarium 1879
sandwich 1762
sandwich toaster 1922
sarrusphone 1856
satellite
  balloon communications 1960
  communications 1958
  Earth Resources Technology (LANDSAT) 1972
  gamma-ray hunting 1973
  geosynchronous 1963
  global communications 1965
  infra-red astronomical 1983
  international ultra-violet explorer 1978
  missile launch warning 1960
  navigation 1963, 1964
  observation/spy 1961
  solar maximum mission 1980
  Sputnik 1957
  tracking and data-relay (TDRS) 1983
  weather 1960
satellite docking 1967
satellite navigator, miniature 1979
satellite tracking of wildlife 1972
satellite transmission, transatlantic 1962
Saturn, observation 1610
saucepan, aluminium 1890
sawing engine 1683
saxophone 1844
scandium 1879
scanner, NMR 1981
scanning disc, Nipkow 1884
'Schaphandre' (diving apparatus) 1855
'Schizophone' 1890
Schottel-Rudder propeller 1950
scissors, mass-produced 1761
scooter 1897
Scotch Tape 1929
Scrabble (game) 1948
screw-cutting lathe 1569
screwdriver 1550
screw-manufacturing machine 1760
screw top (for bottles), external 1852
scribe's outfit 2600 BC
script
  Carolingian miniscule 800
  Demotic 700 BC
sculpture, stone 22000 BC
sea quadrant 1748
seal
  cylinder 3500 BC
  engraved 2400 BC
sealing wax 1556
seamless bust support 1903

seaplane 1910
  flying-boat-launched 1938
  supersonic delta-wing 1954
searchlight, naval 1868
Sears Tower 1973
seasickness preventative 1874
secateurs 1815
secretin 1902
sedan chair 1634
sedimentation equilibrium 1908
seed drill machine 1701
seed-drill plough 85 BC
seismograph
  electromagnetic 1905
  strain 1935
seismometer 1858
selenium 1818
self-raising flour 1871
self-service grocery store 1912
self-starter, automobile 1911
Sellotape 1929
semaphore 1801
  indicator arms 1905
  telegraph 1816
senet (game) 2450 BC
'Sensurround' (cinema) 1974
SEPADS device 1985
Seppings blocks 1800
septic tank 1896
servo-assisted brakes 1919
sewage treatment
  activated sludge 1912
  septic tank 1896
sewage sludge, polymer thickening process 1980
sewers 600 BC
sewing machine 1830
  domestic 1857
  electronic 1975
  four-motion feed 1854
  lock-stitch 1833
sewing needles, bone 17000 BC
sex hormone 1927
sextant 1757
  equatorial 1856
  giant 990, 1425
SGHWR (nuclear reactor) 1968
shaduf (well-sweep) 1400 BC
sheep-shearing machine 1868
Sheffield plate 1743
shell-moulding 1941
shellac mouldings 1868
ship
  all-welded 1919
  armoured 1591
  caloric-engined 1852
  circular 1870
  cupola 1855
  diesel-engined 1904, 1911
  early 8000 BC
  gun-carrying 1339
  ironclad 1787, 1862
  jet-propelled 1839
  merchant 1859, 1862, 1874, 1980
  ocean liner 1881, 1907
  passenger excursion steam-turbined 1902
  rectangular sail 3200 BC
  refrigeration 1880
  research 1950
  steamroller 1896
  steam-turbined 1903
  steel-hulled 1862

Z

Zeppelin raid 1915
zero (mathematics) 10000 BC
Ziggurat, Ur-Nammu 2100 BC
zinc 1541
zinc white 1790
zinc-coated steel 1840

zip fastener 1891
  hook-less 1906
zirconium 1789, 1937
zither, bowed 1823
Zoetrope 180
zoo, public 1793
zoom lens 1932, 1945